Lecture Notes in Physics

Edited by J. Ehlers, München, K. Hepp, Zürich,
H. A. Weidenmüller, Heidelberg, and J. Zittartz, Köln
Managing Editor: W. Beiglböck, Heidelberg

45

Dynamical Concepts on Scaling Violation and the New Resonances in e^+e^- Annihilation

Edited by B. Humpert

Springer-Verlag
Berlin Heidelberg GmbH 1976

Editor
Dr. B. Humpert
Université de Genève
Section de Physique
Départment de Physique Théorique
32, Boulevard d' Yvoy
CH-1211 Genève 4

Library of Congress Cataloging in Publication Data

Humpert, B 1942-
Dynamical concepts on scaling violation and the new resonances in e^+e^- annihilation.

(Lecture notes in physics ; 45)
Bibliography: p.
Includes index.
1. Annihilation reactions. 2. Scaling laws (Nuclear physics) 3. Nuclear magnetic resonance. 4. Hadrons. I. Title. II. Series.
QC794.8.E4H85 539.7'54 75-40317

ISBN 978-3-540-07539-4 ISBN 978-3-540-38061-0 (eBook)
DOI 10.1007/978-3-540-38061-0

Originally published by Springer-Verlag Berlin Heidelberg New York in 1976

DYNAMICAL CONCEPTS ON SCALING VIOLATION AND THE NEW RESONANCES IN e^+e^- ANNIHILATION

by

B. Humpert*

Department of Theoretical Physics
University of Geneva
Geneva - Switzerland

ABSTRACT

We present the essential experimental results on the total and the single-particle inclusive cross sections in e^+e^- annihilation. They show scaling violation and extremely narrow resonances in the energy range beyond 3 Gev. The theoretical ideas about these phenomena are sketched and their characteristics and implications explained. New models are proposed.

*Supported by the Swiss National Science Foundation

C O N T E N T S

1. INTRODUCTION

The recent discoveries of totally unexpected features in e^+e^- - physics have caused a large amount of theoretical speculations, estimations and predictions on the behaviour of electromagnetic interactions at extremely short distances. In this paper(*) we present the essential experimental discoveries - the non-scaling behaviour of $e^+e^- \to$ hadrons[1] and the narrow resonances[2] - and give an introduction to the theoretical explanations of these phenomena.

With the construction of high energy electron accelerators, deep-inelastic electron-proton scattering processes became experimentally possible which give us information how the off-shell photon (exchanged between electron and proton) interacts with the proton. Such experiments permit simultaneous variation of the photon mass $\sqrt{q^2}$ as well as the photon-proton CM-energy $(E_{CM})^2 = W^2$. In particular, in the energy region where $(-q^2, W^2)$ becomes very large, one expects the differential cross section to be only dependent on the dimensionless ratio of these two variables. The experiments in the region $(-q^2, W^2) \lesssim 40$ Gev^2 have sofar confirmed this scaling property[3].

The construction of e^+e^- intersecting storage rings made similar experiments possible however in the time-like region of q^2 . One expected that scaling would also hold in $e^+e^- \to \gamma \to$ hadrons; however, this was not the case. The scaling hypothesis predicts the energy dependence of the total hadronic cross section as : $\sigma(e^+e^- \to h) \propto 1/q^2$, whereas the early experimental results indicated $\sigma(e^+e^- \to h) \cong$ constant beyond 3 Gev. This scaling violation lead to a large amount of theoretical activity which we would like to sketch here[(4)].

A new aspect of this problem recently became apparent when extremely narrow resonances were found at 3.1 Gev and 3.7 Gev. A refined analysis of the experiment then showed more structure around 4.1 Gev and a cross section behaviour between $2.0\ \text{Gev}^2 \leq q^2 \leq 9.0\ \text{Gev}^2$ which is roughly in agreement with scaling[(5)]. It is possible that these two features (scaling violation and narrow resonances) are connected, however, a convincing dynamical mechanism is still lacking.

At the present time, it is difficult to present a conclusive analysis of the theoretical and experimental situation. We therefore sketch the theoretical schemes which suggest violation of scaling or non-asymptotic deviations, as well as the proposed models for the narrow resonances, intentionally not excluding models which, on the basis of the experimental information, seem to be in difficulties or which are based on unusual theoretical assumptions. Our intention is to present some of the proposed explanations, discussing their pros and cons and to maintain interest in ideas outside the main streams.

This paper is divided in three parts. In the first one we present the essential experimental results in e^+e^- annihilation before the discovery of the narrow resonances and subsequently discuss the new experimental data of the narrow resonances. The attempts to understand a possible scaling violation in the present energy range are sketched in the second part. In the third part we discuss the proposed explanations for the newly discovered narrow resonances and wide enhancement.

2. THE EXPERIMENTAL RESULTS

In this section we will present the experimental results of[1, 5, 6]

(1) $e^+e^- \rightarrow$ hadrons : $\sigma_h(q^2)$

(2) $e^+e^- \rightarrow h^\pm + X$: $E\,\frac{d^3\sigma}{dp3}$

(3) $e^+e^- \rightarrow \mu^+ + \mu^-$: $\sigma_\mu(q^2)$

Concerning the first reaction we are interested in the hadronic total cross section. The single-particle inclusive cross section of the second reaction gives us information on the photonic hadron production mechanism. The measurement results of μ-pair production are in agreement with QED calculations apart from the narrow resonances which appear in this channel too[7].

2.1. The total cross section

In the following we will use the notation introduced in Fig. 1. The only invariant variable of this process is the CM-energy square $(E_{CM})^2 = q^2 = (p_+ + p_-)^2$. What is the asymptotic dependence of $\sigma_h(q^2)$ on q^2? For a long time it was a common belief that this quantity would decrease like $1/q^2$.

It was based on the scaling hypothesis which assumes that there is no fundamental scale length in electromagnetic and weak interactions; dimensionality considerations then predict such a characteristic. Early theoretical investigations such as field theory-, dual resonance-, parton- and light-cone-models and a number of other approaches predicted similar decrease. The measurements, extended to $q^2 \lesssim 25$ Gev^2, indicated first an approximately constant behaviour (Fig. 2). Lateron a pronounced resonance structure became apparent (Fig. 3). It is one of the most eagerly awaited answers whether such behaviour will persist. Preliminary results indicate that it falls again for $q^2 > 25$ Gev^2 [13].

These results are often presented as the ratio of the total hadronic to total $\mu^+\mu^-$-production cross sections

$$R(q^2) = \frac{\sigma(e^+e^- \rightarrow h)}{\sigma(e^+e^- \rightarrow \mu^+\mu^-)} \tag{2.4}$$

where the latter is supposed to follow the single-photon exchange approximation with electron and μ-meson point coupling :

$$\sigma(e^+e^- \rightarrow \mu^+\mu^-) = \frac{4\pi}{3} \cdot \frac{\alpha^2}{q^2} \tag{2.5}$$

$R(q^2)$ is represented in Fig. 4 . Instead of a constant behaviour (scaling) the hadron to μ-pair ratio gradually rises beyond $q^2 = 9$ Gev^2 . More refined analysis showed some pronounced resonance structure around $\sqrt{q^2} = 4.1$ Gev followed by a dip around 4.6 Gev and again a rise. The early experimental results had large error bars and globally indicated a rise but gave little indication on the resonance-like structure which now becomes more and more apparent (Fig. 5).

2.2. The inclusive cross section

In a one-particle inclusive measurement the momentum of a chosen type of particle is measured. Such a particle can come from any scattering process : elastic, quasi-elastic, inelastic or deep-inelastic. Its origin is not distinguished.

In the measurements on $e^+e^- \rightarrow h + X$ there was no specific elementary particle chosen, but all charged hadrons instead. Therefore, $h \equiv \{\pi^\pm, K^\pm, p, \bar{p} \ldots\}$. The measured quantity is the sum of the individual cross sections for charged inclusive hadron production. Taking into account the ratio of the average number of produced particles from another measurement $\langle n_{\pi^\pm} \rangle : \langle n_{K^\pm} \rangle : \langle n_p \rangle = 100 : 10 : 1$ one concludes

$$\left(E \cdot \frac{d^3\sigma}{d^3p} \right)_{exp.} = \sum_{i \varepsilon h} E \cdot \frac{d^3\sigma^i}{d^3p} \simeq E \cdot \frac{d^3\sigma^{\pi^+}}{d^3p} + E \cdot \frac{d^3\sigma^{\pi^-}}{d^3p} \tag{2.6}$$

and expects deviations of the order of 10%. Note that the relative inclusive rates are momentum dependent - falling for $\pi^-/h^{(-)}$ and rising for $K^-/h^{(-)}$ respectively $\bar{p}/h^{(-)}$ (Fig. 6).

The most important experimental characteristics are :

i) From simple dimensionality arguments (scaling, section 3.1.) one expects $q^2 \frac{d\sigma}{dx} = f(x,q^2)$, plotted versus $x \equiv 2E/\sqrt{q^2}$, to show no variation with changing q^2 at asymptotic energies ($E \equiv$ energy of inclusive particle). The experiment, however, exhibits a clear q^2-dependence for $x < \frac{1}{2}$ (Fig. 7, 8).

ii) $\frac{1}{\sigma_h} \cdot \frac{d\sigma}{dx}$ is also dependent on x and q^2 (Fig. 9)

iii) The distribution $E \cdot \frac{d^3\sigma}{dp^3}$ versus p is independent of the initial CM-energy q^2 and decreases exponentially with growing momentum, like $\exp(-5 \cdot p)$ (Fig. 10, 11). A similar characteristic was found in hadronic reactions where $E \cdot \frac{d^3\sigma}{dp^3} = (X_{\parallel}, p\ )$ depends on E_{CM} only through the scaling variable $x_{\parallel} \equiv p_{\parallel}/_{p_{max}}$, $p_{max} = \frac{1}{2} E_{CM}$.

The momentum dependence of inclusive (π, K, p) - production is shown in Fig. 12.

iv) The mean momentum per charged particle (hadrons + leptons + ... !) and mean charged multiplicity rise slowly with increasing initial CM-energy : $3.0 \leq \sqrt{q^2} \leq 7.0$ Gev

$$< p_{\pm} > \quad = \quad 0.4 \rightarrow 0.5 \ \text{Gev}/_c \qquad (2.7)$$

$$< n_{\pm} > \quad = \quad 3 \rightarrow 4 \rightarrow \ldots \qquad (2.8)$$

Most recent results indicate $<n_{\pm}>$ to go beyond 4.5 at $\sqrt{q^2} \sim 7$ Gev (Fig. 13).

v) The inclusive angular distribution of the charged particles (expected to be like $(1 + \cos^2\theta)$ in the parton model) is consistent with isotropy for $|\cos\theta| \lesssim 0.6$ and $3.0 \leq \sqrt{q^2} \leq 5.0$ Gev (Fig. 14). The hadronic total - and inclusive cross sections are connected by the energy momentum conservation sum rule

$$P_\mu \cdot \sigma_{tot} = \sum_{all} \int d^3p \cdot p_\mu \cdot \left(\frac{d^3\sigma}{d^3p}\right) \tag{2.9}$$

from which we deduce the relation

$$E_{CM} = \{ \langle n_\pm \rangle \cdot \langle E_\pm \rangle + \langle n_o \rangle \cdot \langle E_o \rangle \} = (\sqrt{q^2}) \tag{2.10}$$

$\langle E_{\pm,o} \rangle$ is the average energy going into one charged/ neutral particle. Experimentally, the first term rises since $\langle E_\pm \rangle$ and $\langle n_\pm \rangle$ rise ; however, its relative contribution $\langle E_\pm \rangle \cdot \langle n_\pm \rangle / E_{CM}$ diminishes. Consequently, more energy goes into neutral particles with growing CM-energy.

(Note $$\langle n \rangle = \sum_h \int_{\frac{2m_\pi}{\sqrt{q^2}}}^{1} \frac{1}{\sigma_h} \cdot \left(\frac{d\sigma}{dx}\right) dx \qquad \left(x \equiv \frac{2E}{\sqrt{q^2}} \right) \tag{2.11}$$

is the charged hadronic multiplicity).

In Fig. 4 we have drawn

$$R_{(\pm)} = \frac{\sigma(\pm)}{\sigma_\mu} , \quad \sigma(\pm) = \frac{\langle n_\pm \rangle \langle E_\pm \rangle}{E_{CM}} \sigma_h \tag{2.12}$$

which indicates the percentage of energy going into charged particles and reflects the non-gegligible amount of energy going into uncharged particles like $\pi^o, K^o, \ldots$, photons, etc. This characteristic is known under the heading "energy crisis". Note that it can also be explained by the rough experimental data and might finally disappear (8).

2.3. The New Resonances

The experimental features just presented are explained by a variety of theoretical schemes which globally can (or cannot) describe the data; they will be presented later. The question of why scaling does not set in as early as in deep inelastic electron proton scattering however is still an unsolved problem. The discovery of extremely narrow resonances in the e^+e^- channel of the reaction $p + Be \rightarrow e^+e^- + X$ at Brookhaven (9) opened another aspect of e^+e^- physics which might be related to the unexpected general characteristics.

2.3.1. The Resonances ψ

Measurements at SLAC (10), ADONE (11) and DESY (12) show resonance spikes in the reactions $e^+e^- \rightarrow$ hadrons and $e^+e^- \rightarrow e^+e^-$, $\mu^+\mu$ at 3095 Mev and 3684 Mev. They were given the names J (Brookhaven), ψ (SLAC) for the first one and ψ' (SLAC) for the second one. The experimental determination of their widths is limited by the beam resolution : $\Gamma_\psi \lesssim$ 1.9 Mev (10).

Gaussian shape distribution function for the energy resolution folded with a Breit-Wigner resonance permits much more precise determination of the actual widths. They were found to be in the range 50 - 100 kev corresponding to a life time of $\tau \sim 10^{-20}$ sec. The cross sections at the top of the resonance $\psi(3095)$ are :

$$\begin{aligned} &\sigma_{\psi}(e^+e^- \to h) \simeq 3000 \text{ nb} && \Gamma_{tot} = 69 \pm 15 \text{ kev} \\ &\sigma_{\psi}(e^+e^- \to \mu^+\mu^-, e^+e^-) \simeq 100 \text{ nb} && \Gamma_{e} = 4.8 \pm 0.6 \text{ kev} \end{aligned} \qquad (2.13)$$

At the resonance $\psi'(3684)$ they are smaller

$$\begin{aligned} &\sigma_{\psi'}(e^+e^- \to h) \simeq 1000 \text{ nb} && \Gamma_{tot} = 225 \pm 56 \text{ kev}^{(13)} \\ &\sigma_{\psi'}(e^+e^- \to e^+e^-, \mu^+\mu^-) \simeq \; ? && \Gamma_{e} = 2.2 \pm 0.5 \text{ kev} \end{aligned} \qquad (2.14)$$

Consequently, the ratio $R \equiv \sigma_h / \sigma_\mu$ jumps from about 3 off the resonance to $R \gtrsim 30$ at the peak of ψ . The shape of ψ is drawn in Fig. 15 and in more details in Figs. 16, 17; similar curves exist for ψ' (Fig. 18). SPEAR has systematically searched for further narrow resonances in the energy range 3.2 - 5.9 Gev by scanning in ~ 2 Mev steps and by imposing $\int \sigma_h(E)\, dE \lesssim 10^3 \text{nb} \cdot \text{Mev}$. No further such objects were found[10d]. The search has been carried out at higher energies and no further narrow resonances were found.[13]

Measurements in the range 2.4 - 5.0 Gev showed a broad enhancement ψ'' with peak at 4.1 Gev (Fig. 3). In the vicinity of this region the cross section varies between 32 nb and 17 nb[10e]. Taking it as a resonance it would have a total width of $\Gamma_{\psi''} \sim 300$ Mev. Table I gives the essential characteristics of the important resonances in e^+e^- channels. The spin-parity of ψ (3.1) was determined by the angular distributions of $e^+e^- \to \mu^+\mu^-$, e^+e^- which are consistent with $J^P = 1^-$. One further knows that the transition : $\psi' \to \psi + \pi^+\pi^-$ exists (Fig. 19). The angular distribution of the $(\psi - 2\pi)$ system is in agreement with the spin-parity assignement $J^P = 1^-$ for ψ'(10f,10h). ψ' and ψ thus seem to carry the same quantum numbers as the photon and might well be directly coupled ' à la vector meson dominance'. Nothing is known sofar about the 4.1 enhancement ($J^P = 1^-$?) which could also be due to a threshold.

2.3.2. The Decay Modes

Some of the essential decay modes of ψ and ψ' and their (preliminary) branching ratios are represented in tables II and III. We mention : a relatively important decay is $2\pi^+ 2\pi^-$+missing mass (π^o?); the $2\pi^+ 2\pi^-$decay is substantially smaller. Off resonance the $2\pi^+ 2\pi^-$fraction is $\sim$ 1/20 of all hadronic decays. On resonance it is estimated by taking $\frac{1}{20}$ Γ $(\psi \to \gamma \to \text{hadrons}) \sim 0.65$ kev and is not very far from the experimental value 0.35 kev. Rough estimates predict 1/3 of all decays via single photon exchange. The 5 pion decay mode seems to be directly coupled. If it occurs via strong interactions, G-parity must be G = -1 and isospin I = 0 or 2 for ψ (3.1), an assignment which is consistent with other observed modes like $\pi^+\pi^- p\bar{p}$, $p\bar{p}$, $\pi^+\pi^- K^+K^-$ and $\Lambda\bar{\Lambda}$. The cross section for an odd number of pions is substantially bigger than that for an even number of pions (5,10g,10i).

No special features are known on the $\psi''(4.1)$-enhancement sofar. Decays like $\psi'' \to \psi' + \pi\pi$ or $\psi + \pi\pi$ have apparently not been detected. The leptonic decay width is a few kev.

2.3.3. Inclusive Distributions

A DASP group has measured the inclusive distribution at ψ (3.1) and preliminary results indicated that it also decreases exponentially, roughly (12e). More recent data from SLAC, which compare the re-scaled inclusive ψ-distribution with the one outside the resonance exhibit deviations of a factor $\sim \frac{1}{2}$ at large x (Fig. 20) (10i). Apart from the difference in the counting rates, all measured quantities such as $\langle p_{\pm} \rangle$, $\langle n_{\pm} \rangle$, etc. on and off the resonance seem to be the same. The ratios $\left(\frac{\pi^-}{h^{(-)}}, \frac{K^-}{h^{(-)}}, \frac{\bar{p}}{h^{(-)}}\right)$ indicate relatively small changes on and off the resonance ψ (3.1) (Fig. 21). However, more recent results on the K^--fraction (Fig. 22) reveal that K^--production is suppressed by $\sim$ 20% at ψ (3.1) . Note that most recent data on particle ratios, in comparison with earlier curves (Fig.21 vs Fig. 6) show a more pronounced decrease in $\pi^-/h^{(-)}$ and increase with $K^-/h^{(-)}$ with growing momentum p (13).

2.3.4. Hadronic Production

$\psi(3.1)$ has been found in the mass spectrum $M^2_{e^+e^-}$ of $p = Be \to e^+e^- + X$ (14a). The initial energy was 28.5 Gev.

The cross section estimate is not without problems since the inclusive distribution in p_t and $p_{//}$ is unknown. The assumption of an exponential cut-off gives :

$$\sigma(p+Be \to \psi+X) \sim 0.1\ ^{nb}/\text{nucleon} \Rightarrow \sigma(p+N \to \psi+X) \sim 1.5\ ^{nb}/\text{nucleon}$$

Apparently, the ψ-enhancement disappears at 20 Gev initial energy (14b). Search for $pp \to \psi' + X$ was negative; this is partially explained by the small branching ratio into $\ell^+\ell^-$ (14c). The background contribution in the 28.5 Gev experiment is small. The resonance to background ratio for e^+e^- - and pp - initiated ψ-production is

$$R(e^+e^-) = 300 \text{ Mev} \ ; \ R(pp) = 1200 \text{ Mev}$$

where we have used the definition : $R(pp) \equiv \left(\frac{\text{Res. - events}}{\text{BG. - events/Mev}}\right)_{pp}$
The Brookhaven ratio is larger than the SPEAR ratio, but in the same order of magnitude (14d,e).
Results with initial neutron energies peaked at 250 Gev show (14f,g)

$$\sigma(n+Be \to \psi+X) \sim 3\ ^{nb}/\text{nucleon} \Rightarrow \sigma(n+N \to \psi+N) \sim 45\ ^{nb}/\text{nucleon}.$$

considerably larger values.

2.3.5. Photoproduction

Search for ψ's produced in photo-initiated reactions initially gave only upper limits of the cross section :

$$E\gamma = 11.1 \text{ Gev} \quad \sigma(\gamma N \to \psi N) \lesssim 1 \text{ nb}^{(15a)}, \ 0.65 \text{ nb}^{(15b)}.$$

$$E\gamma = 18.2 \text{ Gev} \quad \sigma(\gamma N \to \psi N) \lesssim 29 \text{ nb}^{(15c)}, \ 3.7^{+2.2}_{-1.5} \text{ nb}^{(15d)}.$$

Preliminary information from SLAC indicates

$$E\gamma = 17\text{ Gev} \quad \frac{d\sigma}{dt}(\gamma p \rightarrow \psi + X) \propto \cdot e^{(2.6 \pm 0.9)\cdot t}$$

The integrated cross section $\sigma(\gamma p \rightarrow \psi X)$ rises by a factor of 2 in the range 15 Gev $\leq E\gamma \leq$ 18 Gev[15e]. Experiments at FNAL[15f] have discovered a ψ-enhancement and are able to determine the (expected) diffractive shape with an extimated slope around $b \sim 4\text{ Gev}^{-2}$. The cross section value is

$$E\gamma < 200\text{ Gev} \quad \sigma(\gamma + Be \rightarrow \psi + X) = 16 \pm 5\text{ nb/nucleon}$$

giving $\sigma_{tot}(\psi N) \cong 1$ nb. Table VI summarises the parameters of the diffractively produced vector mesons : ρ, ω, ϕ, ψ[15c].

3. THEORIES ON SCALING AND ITS VIOLATION

In this section we present the theoretical developments on scaling violation before the discovery of the new resonances. We start with the definition of scaling and explain its consequences in deep-inelastic electron-proton scattering and e^+e^- annihilation. Subsequently, we present the characteristics of the different approaches explaining scaling violation in e^+e^- annihiliation.

3.1. Scaling - Definition and Hypothesis (16)

If a cross section, for asymptotic values of the relativistic invariant kinematic parameters $\{q_1^2 \ldots q_n^2 | q_\alpha^2 = (p_i + p_j + \ldots)^2\}$ only depends on dimensionless ratios of the latter and no fixed quantities (such as masses, etc.) of the same dimension $(\text{length})^{-2}$ or any other are involved, one calls such property scaling.

The defined scaling property can be considered as a consequence of the scaling hypothesis. Consider a purely leptonic or semi-leptonic reaction - for the latter the sum over all final hadron channels, in addition, is supposed. These can proceed via second order electromagnetic or first order weak interaction. Let $d\sigma$ be the appropriate differential cross section which can, in general, be written as

$$d\sigma = f(s; \{q^2\} ; \{m_\ell^2\} , \{m_n^2\} \ldots) (\text{coupling constant})^2 \quad (3.1)$$

where s is the $(\text{CM-energy})^2$ of the overall process and $\{q^2\} \equiv \{q_1^2,\ldots, q_i^2\}$ represents all further independent invariants that can be formed. $\{m_\ell^2\}$, $\{m_h^2\},\ldots$ stands for the set of lepton and hadron masses and other possible scale parameters with the same dimension (e.g. Regge trajectory slope α', temperature T_o, etc.) which can appear in the description of the process. Leaving the latter aside, the scaling hypothesis states :

i) if $(s, \{q^2\}) \gg \{m_\ell^2\}$ then it is a good approximation to set $\{m_\ell\} = 0$ in the expression for $d\sigma$, and

ii) if $(s, \{q^2\}) \gg \{m_h^2\}$ then it is a good approximation to set $\{m_h\} = 0$ in the expression for $d\sigma$, provided that all final hadronic channels are summed over.

Accordingly, for $(s,\{q^2\}) \gg$ several $(\text{Gev})^2$, one may set as a good approximation $\{m_\ell\} = \{m_h\} = 0$; therefore, eq. (3.1) becomes simply

$$d\sigma = f(s; \{q^2\}) \cdot (\text{coupling constant})^2 \tag{3.2}$$

Apart from the coupling constant, the differential cross section $d\sigma$ now depends only on s and the set of independent kinematical variables $\{q^2\}$ which are the only physical observables with the dimension $(\text{length})^{-2}$ (in natural units $\hbar = c = 1$). All the consequences of the scaling hypothesis can easily be derived by a pure and simple dimensional analysis. The scaling hypothesis means simply the absence of any basic physical energy scale such as $\{m_\ell, m_h\}$ or any other mass. This property enables one to connect various cross sections at a relatively low energy range to those at a much higher energy range.

3.2. Consequences of Scaling

In this section the consequences of the scaling hypothesis in deep-inelastic ep-scattering and e^+e^--annihilation are presented.

3.2.1. Kinematics(17)

We first collect the kinematic relations and formulas for deep inelastic ep scattering using the notation defined in Fig. 23 .

$$q = k-k', \; Q^2 = -q^2, \; m\nu = pq = m(E-E')$$
$$W^2 = (p+q)^2 = 2m\cdot\nu+m^2 - Q^2, \; \cos\theta_e = \vec{k}\cdot\vec{k}' \tag{3.3}$$
$$\omega = \frac{2m\nu}{-q^2} \; \left(= \frac{1}{\xi}\right) = \frac{-q^2 - m^2 + W^2}{-q^2} \cong 1 + \frac{W^2}{Q^2}$$

m is the target mass (here $m = M_p$), $k^\mu = (E,\vec{k})$ and $k'^\mu=(E',\vec{k}')$ are the initial and scattered 4-momenta of the electrons in the lab-frame and $1 \leq \omega \leq \infty$ is the scaling variable. Under the assumption of single photon exchange the inclusive cross section factorizes in a lepton part $L^{\mu\nu}$ and a hadron par $W^{\mu\nu}$

$$E' \cdot \frac{d\sigma}{d^3k'} = \frac{4\alpha^2 m^2}{Q^4 E} \cdot L\mu\nu \cdot W^{\mu\nu} \tag{3.4}$$

Due to the optical theorem (Fig. 24) $W^{\mu\nu}$ may be related to the off-shell Compton amplitude by

$$W^{\mu\nu}(p,q) = \sum_x \langle p|j^\mu(0)|x\rangle\langle x|j^\nu(0)|p\rangle\ \delta^4(p+q-p_x)$$
$$= (-g^{\mu\nu}+\frac{q^\mu q^\nu}{q^2})\cdot W_1 + \frac{1}{m^2}(p-\frac{pq}{q^2}\cdot q)^\mu(p-\frac{pq}{q^2}\cdot q)^\nu\cdot W_2 \quad (3.5)$$

The inclusive cross section then reads

$$E'\frac{d^3\sigma}{d^3k'} = \frac{4\alpha^2E'}{Q^4}\{2\cdot W_1(\nu,Q^2)\cdot\sin^2\frac{\theta}{2} + W_2(\nu,Q^2)\cdot\cos^2\frac{\theta}{2}\} \quad (3.6)$$

It contains explicitly the two structure functions W_1 and W_2 which describe the process : $\gamma(q^2) + p \to$ hadrons. They have the dimension $[1/M]$. The dimensionless structure functions mW_1, νW_2 were found to be more convenient for the description of this process and thus contain all information about it.

The description of the e^+e^--annihilation process is analogous (Fig. 1).

$$q = (q_+ + q_-)\ ,\ s = q^2 = E_{CM}^2\ ,\ m\nu = -p_h\cdot q = -\sqrt{q^2}\cdot E$$
$$W^2 = (q-p_h)^2 = 2m\nu + m^2 + q^2\ ,\ \cos\theta_h = \vec{q}\cdot\vec{p} \quad (3.7)$$
$$x\ (\equiv\omega) = \frac{2m\nu}{-q^2} = \frac{2E}{\sqrt{q^2}} = \frac{E}{E_{Max}}\ ,\ E_{Max} = E_{CM}/2\ .$$

m is the inclusive particle's mass, $p_h = (E,\vec{p})$ its center-of-mass 4-momentum and $0 \le x \le 1$ is the physical range of the scaling variable. Note in ep scattering

$(\nu, -q^2) > 0$ whereas $(\nu, -q^2) < 0$ in e^+e^- annihilation. The kinematical regions are represented in Fig. 25 . The inclusive cross section may be determined as in the deep-inelastic case (since they are one and the same process in different kinematic regions) :

$$\left(E\cdot\frac{d^3\sigma}{d^3p}\right) = \frac{2m\alpha^2}{q^4}\left\{ 2\cdot\bar{W}_1(\nu,q^2) + \left(\frac{\nu^2}{q^2} - 1\right)\cdot\bar{W}_2(\nu,q^2)\cdot\sin^2\theta_h \right\} \tag{3.8}$$

The two structure functions $m\bar{W}_1$ and $\nu\bar{W}_2$ reflect the dynamics of single photon annihilation. They are defined again through the invariant expansion of virtual Compton scattering

$$\begin{aligned} W^{\mu\nu}(p,q) &= \sum_X \langle 0|j^\mu(0)|h,X\rangle\langle h,X|j^\nu(0)|0\rangle\, \delta^4(q-p_h-p_X) \\ &= \left(-g^{\mu\nu} + \frac{q^\mu q^\nu}{q^2}\right)\cdot\bar{W}_1 + \frac{1}{m^2}\cdot\left(p_h - \frac{p_h\cdot q}{q^2}\right)^\mu\left(p_h - \frac{p_h\cdot q}{q^2}\right)^\nu\cdot\bar{W}_2 \end{aligned} \tag{3.9}$$

3.2.2. $e^-p \to e^- + X$

According to the scaling hypothesis stated above, one should expect that $m\bar{W}_1$, $\nu\bar{W}_2$ are functions of the ratio $W^2/-q^2$ only for large values of the invariant variables. The experiment was performed such that the quantity

$$\omega \simeq 1 + W^2/Q^2 \;,\; 1 \leq \omega \leq \infty \tag{3.10}$$

was kept fixed as a parameter whereas (Q^2,W^2) were given large values.

$$\underline{\text{Bj-lim}} : \quad (Q^2, W^2) \to \infty$$
$$\omega = 1 + \frac{W^2}{Q^2} = \text{fixed} \tag{3.11}$$

Such a limit was investigated by Bjorken in field theory and lead to the Bjorken scaling law[18] :

$$\text{Bj-lim} \quad m_p W_1 \; (q^2, W^2 \; ; \; m_h^2) \to F_1(\omega)$$
$$\text{Bj-lim} \quad \nu \cdot W_2 (q^2, W^2 \; ; \; m_h^2) \to F_2(\omega) \tag{3.12}$$

The experimental results sofar are consistent with the scaling hypothesis although there are indications of a small violation in $\mu p \to \mu + X$[3,19].

3.2.3. $e^+e^- \to$ hadrons

How well does this hypothesis work for the hadronic total cross section in e^+e^- annihilation ? The only relativistic invariant variable here is $q^2 = (q_- + q_+)^2 \geq 0$. Since we know that this process is of electromagnetic nature the total cross section must be proportional to α^2 . Its dimension is $(\text{length})^{-2}$; therefore, if scaling is correct, it can only have the form :

$$\sigma(e^+e^- \to h) \overset{q^2 \to \infty}{\simeq} \text{const} \; \frac{\alpha^2}{q^2} \qquad \text{(Scaling)} \tag{3.13}$$

for asymptotic values of q^2 .

At relatively low energies, $q^2 \leq 9 \text{ Gev}^2$, the experimental results do not show disagreement with scaling behaviour (Fig.5)

This statement has to be taken with care. The region $q^2 \leq 3$ Gev^2 is dominated by ρ , ω, ϕ, ϕ' - thus non asymptotic. The curve in the region $3 \leq q^2 \leq 9$ Gev^2 seems to agree rather well with the scaling hypothesis. However, there still might exist resonances which add up to a smooth behaviour and leave a falsified overall picture. The early results at higher energies indicated

$$\sigma(e^+e^- \to h) \underset{\smile}{\overset{9 \leq q^2 \leq 25 \text{ Gev}^2}{}} \text{const} \tag{3.14}$$

One now knows that there is structure around 4.1 Gev due to resonances or thresholds and it is not impossible that such behaviour will persist.

At the present time it is therefore difficult to decide whether there is finally scaling violation or not since the asymptotic region has not yet been reached. Large masses appear in the theory which can no longer be neglected. Thus, one of the premises of the scaling hypothesis is no longer satisfied, namely : $q^2 >> \{m_h^2\}$.

We mention that the integrated cross section for $\mu^+\mu^-$-production in e^+e^- annihilation can be obtained by the same chain of arguments and agrees with QED single-photon exchange which also fixes the constant :

$$\sigma(e^+e^- \to \mu^+\mu^-) = \frac{4\pi}{3} \frac{\alpha^2}{q^2} \tag{3.15}$$

The experimental results from SPEAR [7] are in agreement with this form. According to the scaling definition, the large ψ-mass should here destroy this property. It only appears in higher order hadronic corrections and thus leads to negligible effects. The measurement results also permitted new tests on the validity of QED through the cutoff parameter $\Lambda \gtrsim 20$ Gev in

$$F(q^2) = 1 \pm q^2/(q^2 \pm \Lambda^2) \tag{3.16}$$

It was introduced to parametrize an eventual modification of single photon exchange due to a modified photon propagator, heavy neutral bosons or other QED breakdowns ( ). It is now obvious that scaling - stated differently - means

$$R(q^2) \equiv \frac{\sigma(e^+e^- \to h)}{\sigma(e^+e^- \to \mu^+\mu^-)} \xrightarrow{q^2 \to \infty} \text{const.} \tag{3.17}$$

where "const" is fixed in specific models.

3.2.4. $e^+e^- \to h + X$:

Supposing inclusive e^+e^- annihilation proceeds via single-photon exchange (Fig. 1) one notices immediately the similarity to the process $e^+e^- \to \bar{p} + X$ which is obtained by crossing from $e^-p \to e^-X$.

$$\boxed{e^+e^- \to \pi^\pm + X} \longleftrightarrow e^- + \pi^\pm \to e^- + X$$

$$\uparrow$$

$$e^+e^- \to \bar{p} + X \longleftrightarrow \boxed{e^- + p \to e^- + X}$$

The measured reactions above are inclosed in boxes. Inclusive π-production in e^+e^- annihilation corresponds to deep inelastic $e\pi$ scattering. The realization of such experiment seems to be difficult. However, deep-inelastic scattering : $\mu + \pi \to \mu + X$ will become possible in the near future through the reaction : $\mu + p \to \mu + N^* + X$ (Fig.26) [20a]. The invariant variable $t = (p - p_1)^2 < 0$ can be adjusted at will and is chosen at $t = 0$ which is very close to the pion pole. There are a number of further diagrams contributing to this process such as nucleon pole contributions, A_2 exchange instead of π, etc. which, however, can all be eliminated because of their different kinematical characteristics [20b,c].

The kinematic description of inclusive e^+e^- annihilation (with single photon exchange) is essentially as in deep inelastic ep scattering (as we have seen above). The experiment was performed again with the ratio W^2/q^2 = fixed by the variable

$$x \equiv \frac{2E}{\sqrt{q^2}} \simeq 1 + \frac{W^2}{-q^2} \quad (0 \leq x \leq 1) \tag{3.18}$$

which determines how much of the initial energy $E_{Max} \equiv E_{CM}/2 = \frac{1}{2}\sqrt{q^2}$ goes to the inclusive particle. The differential cross section (instead of $d\sigma/dE_h$) has the form :

$$\frac{d\sigma}{dx} = f \cdot \left[2m\bar{W}_1 + g \cdot \nu\, \bar{W}_2\right] \tag{3.19}$$

$$f = f(q^2, W^2) \equiv \frac{2\pi\alpha^2}{q^2}\, x \cdot \left(1 - \frac{4m^2}{q^2x^2}\right)^{\frac{1}{2}}$$

$$g = g(q^2, W^2) \equiv \frac{1}{3} \cdot \left(1 - \frac{4m^2}{q^2x^2}\right)$$

Identifying $m \equiv m_\pi$ and taking the Bjorken limit (with $q^2 > W^2 > o$ large) reveals :

$$q^2 \cdot \frac{d\sigma}{dx} = \text{fct.}\ (x,\ q^2) \to f(x) \tag{3.20}$$

is dependent on the initial energy $\sqrt{q^2}$ only through the scaling variable x if the scaling hypothesis in the case of annihilation is correct :

$$\begin{aligned} &\text{Bj-lim}\ \ m\,\bar{W}_1\ (q^2, W^2;\ m_h^2) \to \bar{F}_1(x) \\ &\text{Bj-lim}\ -\nu\,\bar{W}_2(q^2,\ W^2;\ m_h^2) \to \bar{F}_2(x) \end{aligned} \tag{3.21}$$

Eq. (3.20) could also have been derived by simple dimensional analysis.

In conclusion, one expects that $q^2 \cdot \frac{d\sigma}{dx}$ shows no dependence on q^2 for fixed x . This, as we have seen in section 2.2., is just not satisfied or only for $x > \frac{1}{2}$. Therefore, for $x < \frac{1}{2}$ there is a scaling violation.

3.2.5. e^-p versus e^+e^-

We add in this section the modelling attempts which assume that the hadron creation process is essentially the same in e^+e^- annihilation and deep inelastic ep scattering. Thus, scaling holds for both but is hidden at present energies in the former reaction because of threshold effects and the creation of large mass systems.

However, let us first characterize the differences between inclusive e^-p and e^+e^- experiments. The inclusive particle in the former is a proton with $m_p \sim 1$ Gev, whereas in the latter it is the π^+-meson with a mass which is roughly 10 times smaller. The mass of the off-shell photon is below zero in ep scattering whereas it is above zero in e^+e^- annihilation. In the complex q^2-plane one expects thresholds and cuts along the positive real q^2-line due to particle creation whereas there are none along the negative real axis (Fig. 27). In both reactions W^2 is positive and large.

Neglecting these differences, the hypothesis has been put forward that there is scaling in both reactions whose matrix elements are then connected by analytic continuation from $q^2 < 0$ to $q^2 > 0$. The structure functions F_1, F_2 of models in deep inelastic ep scattering are continued into the region of e^+e^- annihilation using the Gribov-Lipatov relations

$$\bar{F}_2(x) = \frac{1}{x} F_1 \left(\frac{1}{x}\right)$$
$$\bar{F}_2(x) = \frac{1}{x^3} F_2 \left(\frac{1}{x}\right) \tag{3.22}$$

which were found in perturbative models of cut-off field theories (21, 22). The fact that eqs. (3.22) relate data for two kinds of processes makes it interesting to test these relations. A comparison of $e^+e^- \to \bar{p} + X$ with $e^- + p \to e^- + X$, using the presently available data supplemented by some simplifying assumptions, shows : the reciprocity relation

$$\frac{x}{\sigma_\mu} \frac{d\sigma}{dx} = x^3 \cdot \bar{F}_2 = F_2\left(\frac{1}{x}\right) \tag{3.23}$$

is in fairly good agreement with the data for these two processes (Fig. 28). Applying the analogous arguments to $e^+e^- \to \pi + X$ with the annihilation data as input, one predicts the deep inelastic structure function

$$F_2^\pi \sim 10 \cdot F_2^p \quad \text{at} \quad x = 2 \tag{3.24}$$

This is unreasonably large, reflecting the fact that σ_h (whose main contribution comes from π-production : $e^+e^- \to \pi + X$) is large and violates scaling! (3,23)

In e^+e^- annihilation scaling is seen much later because of the constant opening of new channels which give rise to considerable correction effects. We mention particular models.

One might assume that the inclusive distribution of baryons (and anti-baryons) $e^+e^- \to h_B + X$ does scale. Subsequent annihilation of $B\bar{B}$-pairs create many π-mesons (by final state interaction) which overshadow the scaling behaviour (Fig. 29)(24). The total cross section in such a model is determined by the energy conservation sum-rule. The (scaling) inclusive distribution for $B\bar{B}$-production is determined by the structure functions of deep inelastic ep scattering, analytically continued via Gribov-Lipatov relations to the e^+e^- annihilation region. The inclusive cross section for π-mesons coming from annihilating $B\bar{B}$-pairs is then equal to the product of the $B\bar{B}$-production cross section times the π-decay distribution of a $B\bar{B}$-system at an average invariant mass $\bar{M}$ which is adapted from a statistical model for $p\bar{p}$-annihilation (Fig. 30)

$$E\cdot\frac{d^3\sigma}{d^3p} = \{ R_B(q^2) + R_m(q^2) \}\cdot\sigma_\mu(q^2)\cdot E\,\frac{d^3\sigma}{d^3p}\,(\bar{M}^2) \qquad (3.25)$$

Arguments in similar direction assume the inclusive creation of intermediate baryon and meson resonances to scale (Fig. 31)(25). Inputs of this model are the Callan-Gross relation and a dual-model ansatz for the structure function which was obtained in investigations for the deep inelastic region and analytically continued via Gribov-Lipatov reciprocity relations to the process $e^+e^- \to h + X$. Contributions due to anomalous singularities which appear after the crossing procedure are here also taken into account and turn out to give substantial contributions. This model fits surprisingly well the inclusive distribution and predicts a rising total cross section which, however, is below the measured one because of the large number of produced neutral particles (Fig. 32).).

We sketch some new arguments which are based on the quasi two-body reaction : $e^+e^- \to h_1 + h_2$ (26a). A significant fraction of the total cross section in $\sqrt{q^2} \leq 3$ GeV can be reproduced if (h_1, h_2) are any pair of mesons (or-resonances) in the lowest super multiplet of SU(6) $(\pi, \rho, K, K^* , ...)$ and if one supposes that quasi two-body has an important rule to play production, in particular πR, in the extended region $\sqrt{q^2} \leq 5$ GeV. Let us consider in more detail the inclusive cross-section for the reaction : $e^+e^- \to \bar{\pi} + R$ (R ≡ arbitrary particle of spin J_R). By introduction of singularity and contraint free amplitudes and closer inspection of the matrix element, the distinctive form

$$q^2 \cdot \frac{d\sigma}{dx} \sim x^{2J_R} \cdot (q^2)^{2\,J_R - 1} \cdot |F_o(J_R, q^2|^2 \qquad (3.26)$$

is found, where F_o is the form factor. It is supposed (in agreement with other experimental information)[26b] to decrease universally for all resonances R like

$$F_o(J_R, q^2) \underset{}{\overset{q^2 \to \infty}{\sim}} (q^2)^{-1-J_R} \tag{3.27}$$

The arguments now go as follows : At large q^2 the inclusive cross section falls to zero like $\sim (q^2)^{-3}$ in roughly the same manner for all resonances. However, for q^2 near thresholds W_R^2 ($x \to 0$) the cross section behaviour is governed by threshold factors which are strongly dependent on J_R. Insofar as one expects high spins to go with higher masses, then the rise from threshold is steeper for larger masses W_R^2 (viz. larger J_R) and hence, for fixed $x \cong 1 + \frac{W_R^2}{-q^2}$, with higher q^2 (Fig. 33). To say it simply, the apparent scaling violation can, according to this picture, be associated with thresholds opening up. Deep inelastic scattering also fits nicely into this scheme. To simplify the presentation let us go along a line with W^2 = fixed as in Fig. 25. The crucial difference between space-like and time-like q^2 is illustrated in Fig. 34. In $e^+e^- \to \bar{\pi} + R$, the physical region thresholds are always present and lead to small x behaviour as explained above and hence are responsible for a seeming violation of scaling. In $e^- \pi \to e^- R$ the threshold goes further and further away from the physical region as the state h_2 increases in mass (since we move upwards on the line W^2 = fixed) and scaling improves. The "symmetry point" between the space like and time-like region is seen to be $q_o^2 = (m_\pi^2 + m_R^2)$ and not $q^2 = 0$ as often imagined.

Decisive theoretical predictions of these approaches seem quite difficult. Experimental analysis of detailed channels certainly will help to clarify. Some eventwise jet-structures, in particular kinematical regions, as well as threshold enhancements might be seen or washed out. Scaling in inclusive production through large mass intermediate particles (supposed to scale) is falsified through kimematical-, threshold-, decay- and other factors and therefore can hardly be tested in the present energy range.

3.3. Models : $\gamma \to$ hadrons

In the previous sections our main interest was : scaling- or violation ? Here we present the modelling attempts under the more general heading of "photonic hadron creation", since the asymptotic region, where scaling is experimentally verified, seems further out than expected. In addition, new resonances ψ are now known to exist whose decay mechanisms are unclear and might be different from those of the photon. Most of the proposed models have enough flexibility to accommodate a behaviour conform with scaling or violation in the present energy range.

The remainder of this section is organized as follows : since all models assume the creation of sub-constituents out of the vacuum, we first briefly repeat some "folklore" on Dirac's vacuum, in order to facilitate intuitive understanding of the models. Subsequently we present the various modelling attempts to explain hadron production in e^+e^- annihilation and discuss some of their characteristics.

3.3.1. Physical Vacuum[27]

Let us consider the energy spectrum of a charged Fermion which is described by the Dirac equation. It can have positive energy $+E_o$ (particle) or negative energy $-E_o$ (anti-particle) (Fig. 35). In Dirac's interpretation the anti-particle is considered to be the missing particle in the completely filled negative energy spectrum and thus appears as a particle with opposite charge. The necessary energy to create a particle-antiparticle system comes from its surrounding. The specific form of this energy transmission remains unspecified (Fig. 36). One might suppose the vacuum consists of a large number of such particle-antiparticle pairs (which might be identified with partons, quarks, etc.), in a constant process of association and dissociation, such that it may be considered as a fluctuating gas or liquid of quasifree $q\bar{q}$-pairs. This gas then has a certain temperature; for $T = 0$ no excitations exist and all negative energy states are occupied. With increasing temperature the number of excited pairs, and thus their density, increases.

We now present the models on $e^+e^- \to$ hadrons.

3.3.2. Partons [28]

One of the most useful models to describe and connect hadron production in deep-inelastic ep -scattering and e^+e^- annihilation is doubtless the parton model.

The experimentally confirmed scaling behaviour in deep-inelastic ep-scattering led to the hypothesis that hadrons are composed of sub-particles, called PARTONS ; they are supposed to be quasi-free and with limited transverse momentum within the hadron (Fig. 37). The interaction with the highly virtual electromagnetic current, assumed to be incoherent, is described in impulse approximation in an infinite momentum frame. Each type (i) of partons with charge Q_i carries a fraction ξ_i (with probability $f(\xi_i)$) of the nucleon's total longitudinal momentum : $P_z = \xi_i \cdot p_{z,i}$. The structure functions then are (in the scaling region) :

$$\begin{aligned} \nu W_2(\nu,q^2) &\to F_2(\xi) = \sum_i Q_i^2 \cdot \xi \cdot f_i(\xi) \\ m_N \cdot W_1(\nu,q^2) &\to F_1(\xi) = \sum_i Q_i^2 \cdot f_i(\xi) \end{aligned} \tag{3.28}$$

where ξ is the fractional momentum of the struck parton as well as the value of the scaling variable $\frac{1}{\xi} = \omega \equiv \frac{2m\nu}{-q^2}$
Not much is known about the dynamical properties of the partons. The questions are still open (although a number

of solutions have been proposed) of why individual partons do not seem to come out and appear as individual objects and of which forces hold the compound of partons together. Such a scheme with the additional assumption PARTONS ≡ QUARKS, gave a number of predictions which are in quite good agreement with the experiment. The spin ½ nature of the partons gives: $2\xi \cdot F_1(\xi) = F_2(\xi)$. The valence quarks p,n,λ in the nucleon are responsible for the non-diffractive contribution. It is parametrized by the distribution functions $p(\xi), n(\xi), \lambda(\xi)$ and their anti-partners $\bar{p}(\xi)$,... . A (substantial) diffractive contribution is assigned to the isoscalar sea of quark-antiquark pairs within the nucleon. Such framework permits description and connection of all experimental results in deep inelastic lepton-nucleon scattering[28,3].

The parton model was taken over to e^+e^- annihilation. There exists a variety of different formulations of one and the same physical idea, each in its own language :

- phenomenological parton models[29]
- covariant parton model[30]
- field theoretic parton models[31]
- bag models[32]

It is not our intention to compare or to explain these attempts here[33]; we rather explain the measurable consequences of this picture which apply to all formulations.

The photon is assumed to create by point-interaction a parton-antiparton pair ($q\bar{q}$-pair) out of the vacuum which subsequently pulls out further $q\bar{q}$-pairs and thus creates hadrons (Fig. 38) . Since the variable $x = \frac{2E}{\sqrt{q^2}} \cong \frac{p}{p_{max}}$ expresses the percentage of parton momentum which is transmitted to the inclusive hadron, $q^2 \frac{d\sigma}{dx}$ may be interpreted as the probability for producing a hadron with momentum fraction x (Fig. 39). Transverse momentum cut-off is supposed.

Inherent properties of this scheme are [4] :

i) For asymptotic q^2-values $\sigma_h = \frac{4\pi\alpha^2}{3} (\sum_i Q_i^2) \cdot \frac{1}{q^2}$ scales. Stating this another way, one may say the asymptotic hadron to μ-pair cross section ratio approaches a constant value :

$$R_\infty = \sum_{\frac{1}{2}} Q_i^2 + \frac{1}{4} \sum_{0} Q_i^2 \qquad (3.29)$$

The value of the constant depends on the supposed higher symmetry structure of the partons which are identified with quarks. Scaling is expected to set in late.

ii) The parton spin determines the inclusive angular distribution; it is expected to be

$$\frac{d\sigma}{d\cos\theta} \propto (1+\cos^2\theta) \cdot \sum_{\frac{1}{2}} Q_i^2 + \tfrac{1}{2}(1-\cos^2\theta) \cdot \sum_{0} Q_i^2 \qquad (3.30)$$

and should show eventwise jet structure. Due to the parton's large momentum, one expects that hadrons, created by the $q\bar{q}$-pair, are emitted in the direction of q and $\bar{q}$ (Fig. 40). A simple model, with a hadron momentum distribution of the form of a cigar shows that this assumption leads to a scaling violating correction term which is strongly felt at small x :

$$q^2 \cdot x \cdot \frac{d^2\sigma}{dx\, d\cos\theta} = \{ (1+\cos^2\theta) + \frac{f(\cos\theta)}{x^2 \cdot q^2} \} \tag{3.31}$$

iii) The inclusive cross section approaches its scaling limit, as we have just seen, very slowly - in particular at small x-values.

$$q^2 \cdot \frac{d\sigma}{dx} \rightarrow f(x) \ , \quad x \equiv \frac{2E}{\sqrt{q^2}} \tag{3.32}$$

iv) The inclusive cross sections for π^o, π^+ or π^- production should be equal and π-mesons should predominantly be produced.

v) The mean multiplicity $< n >$ is expected to rise logarithmically and in consequence $< p >$ should rise linearly with $\sqrt{q^2}$.

Despite its simplicity and attractive features, most predictions of the parton model do not seem to receive support from the experiment in the range $q^2 \lesssim 25\ \mathrm{Gev}^2$. A number of modifications have therefore been proposed.

a) Partons with Structure [34]

Partons might have a substructure and not interact "point-like" with the off-shell photon as supposed in deep inelastic ep scattering. This is described by a form-factor $F(q^2) \sim (1 - q^2/\Lambda^2)^{-1}$ and an anomalous magnetic momentum μ_Q and leads to the following modifications : at extreme energies scaling is violated as

$$R(q^2) = \{1 + \frac{1}{2}\mu_Q^2 \cdot q^2\} \cdot |F(q^2)|^2 \ (\sum_i Q_i^2) \qquad (3.33)$$

The form-factor mass and the anomalous magnetic momentum of the partons are adjusted to fit the present experimental data : $\Lambda \sim 10$ Gev , $\mu_Q \sim 0.02\ \mathrm{Gev}^{-2}$. Thus R rises up to $q^2 \sim 100\ \mathrm{Gev}^2$ and then rapidly decreases like $1/q^2$. The inclusive distribution $q^2 \frac{d\sigma}{dx}$ shows scale-breaking effects at small x due to the anomalous magnetic moment contribution (and the form-factor) which also is responsible for a change in the angular distribution :

$$\frac{d\sigma}{d\cos\theta} = \left[(1+\cos^2\theta) + \mu_Q^2 \cdot q^2\ (1-\cos^2\theta)\right] \cdot \mathrm{fct.}\ (q^2) \qquad (3.34)$$

In the present energy range it is almost isotropic. $\frac{1}{\sigma} \cdot \frac{d\sigma}{dx}$ is expected to scale - in e^+e^- annihilation as well as in deep-inelastic ep scattering. Sensitive tests of this scheme are : in e^+e^- annihilation the mean energy square of the secondaries, which emphasizes the large x-region, grows as $< E_\pi^2 > \sim q^2$. In polarized electroproduction certain structure functions must have an order of magnitude deviation; in addition νW_2 will scale whereas a considerable scaling violation is expected in W_1 - a reflection of the rising σ_L/σ_T -ratio. There should be large scale-breaking effects in neutrino scattering, but not for antineutrinos.

b) Spin-1 Partons (35)

Spin-1 partons besides spin-½ (and spin-0) ones have been proposed with an electric quadrupole momentum coupling neglecting all other possible couplings. The cross section

$$\sigma_h = \frac{\pi\alpha^2}{3\cdot M_V^2}\{1 - \frac{M_V^2}{q^2} - \frac{12M_V^2}{q^4}\}\sqrt{1 - \frac{4M_V^2}{q^2}} \rightarrow \frac{\pi\alpha^2}{3M_V^2} \qquad (3.35)$$

then approaches a constant value which is dependent on the mass M_V of the vector partons ; thus $R_\infty = q^2/4M_V^2$. The inclusive distribution, expected to be changed at small x , depends on how the charged vector partons evolve into hadrons. A jet-like structure should again be seen with an angular distribution of the jets as for spin - ½ partons : $(1 + \cos^2\theta)$ (35a). The mean multiplicity for production of spin-1 hadrons approaches a constant value ; in the case of spin-1 partons it grows logarithmically. Scaling violation in deep inelastic ep scattering is hidden at present energies. Such models, however, are only renormalizable if $R(q^2)$ is asymptotically constant - a possibility for mesons with Yang-Mills couplings. In such theories (with symmetry $SU_2 \times U_1$) neutral and charged spin-½ partons as well as charged spin-1 partons are assumed to exist. The asymptotic cross sections scale

$$\sigma(e^+e^- \to q\bar{q}) = \frac{\pi\alpha^2}{3q^2} \quad , \quad \sigma(e^+e^- \to \rho^+\rho^-) = \frac{\pi\alpha^2}{12\,q^2} \tag{3.36}$$

and the hadron to μ-pair ratio approaches $R_\infty = 9/16$. Scaling is achieved by form-factor-like modifications of the vertex functions.

c) New Degrees of Freedom (4)

Partons with different higher symmetry properties might be created. If they are supposed to have an additional degree of freedom beside of their usual SU_3-structure (eg. colored quarks), a step-like increase in $R(q^2)$ is explained by its excitation. The same effect should show up if new types of constituents (eg. charmed quarks) with large masses are created out of the vacuum. The asymptotic value R_∞ then essentially depends on the higher symmetry structure of these constituents. Typical values for colored quark schemes are :

$R_\infty = 2/3$ (G-Z quarks) $= 10/9$ (... + charmed quark)

$= 2$ (G-Z quarks + colour) $= 10/3$ (... + charmed quark)

$= 4$ (H-N quarks + colour) $= 6$ (... + charmed quark)

The addition of a charmed quark changes these values (see second column). Other effects are expected which will be discussed in chapter 4.

3.3.3. Vector Mesons [36]

In this section we follow ideas which assume that the hadronic final state system is created by the off-shell photon through the intermediary of hadronic single particle states. The q^2-dependence of the matrix element is thus dominated by resonance poles.

The vector meson dominance model (VMD) was quite successful in describing high energy photoproduction of quasi two-body final states. Such a process exhibits a hadron-like behaviour of the photon which was well explained by this model. It assumes, loosely speaking, that the photon couples directly to the vector meson resonances $R \equiv (\rho,\omega,\phi)$ which then undergo a strong scattering process (Fig. 41). Such a picture also received support from the measurements in e^+e^- annihilation where these resonances appeared in exclusive π- and K-channels (Fig. 42), and the fact that the resonance form of the total cross section [37]

$$\sigma_h = 16\pi^2\alpha^2 \sum_{R_n} \frac{1}{f_n^2} \frac{m_n \cdot \Gamma_n}{(q^2-m_n^2)^2 + m_n^2 \cdot \Gamma_n^2} \qquad (3.37)$$

represented well the data in the low energy region. f_n is the photon-vector meson coupling constant : $e \cdot m_n^2/f_n$, with

$$\Gamma(R_n \to e^+e^-) = \frac{\alpha^2}{3} \frac{m_n}{(f_n^2/4\pi)} \tag{3.38}$$

In some models f_n is assumed to be q^2-dependent, in others it is constant.

The discovery of higher mass resonances (ρ',ρ'',..., etc.) and the dual resonance model suggested that there might be an infinity of such objects with a linear mass spectrum : $m_n^2 = m_o^2 \cdot (1 + a \cdot n)$ [38] (Fig. 43). Choosing in this "extended vector meson dominance model" (EVMD) the coupling f_n and width Γ_n appropriately, one can generate any asymptotic behaviour : decreasing, constant or increasing. We distinguish the following cases :

i) VMD : The cross section is supposed to be dominated by the lowest resonances only

$$\sigma_n \to \frac{16\pi^2\alpha^2}{q^4} \left(\frac{m_n \cdot \Gamma_n}{f_n^2}\right) \tag{3.39}$$

ii) EVMD (scaling) : Adjusting the coupling constants and widths as [36b]

$$\frac{\Gamma_n}{m_n} = \frac{\Gamma_\rho}{m_\rho} = \gamma = \text{const.} , \quad \frac{m_n^2}{f_n^2} = \frac{m_\rho^2}{f_\rho^2} = b = \text{const.} \tag{3.40}$$

the total cross section scales

$$R_\infty = \frac{16\pi}{m_\rho^2} \frac{b^2}{a} \{\tfrac{1}{2}\pi + \text{arctg}(\frac{1}{\gamma})\} \simeq \frac{8\pi^2}{f_\rho^2} \sim 2.5 - 3 \tag{3.41}$$

One might wonder whether an infinity of narrow resonances like ψ , ψ',... leads to a considerable modification. An estimation[39] shows that it would bring another factor of 1.2 leading to

$$R_\infty = 2.5 + 1.2 = 3.7 \tag{3.42}$$

Note that these values depend on the choice of a ; the above value is for a = 2 .

iii) EVMD (non-scaling)[38b]: If the widths and coupling constants are q^2-dependent, the constraint

$$\frac{\Gamma_n(q^2)}{\Gamma_\rho} = \left(\frac{q^2}{m_\rho^2}\right)\cdot\frac{m_n}{m_\rho} \quad , \quad \frac{f_n(q^2)}{f_\rho} \simeq \frac{q^2}{4m_\rho^2} \tag{3.43}$$

leads to a constant hadronic total cross section

$$\sigma_h \simeq \frac{1}{2}\left(\frac{\pi\alpha}{m_\rho \cdot f_\rho}\right)^2\left\{\mathrm{arctg}\,\frac{m_\rho}{\Gamma_\rho} + \frac{\pi}{2}\right\} \tag{3.44}$$

A scaling decay distribution $R_n \rightarrow$ hadrons

$$E\cdot\frac{d^3(m_n\Gamma_n)}{d^3p} \propto \frac{f_h(x)}{x^2} \tag{3.45}$$

then reveals for the inclusive single particle distribution $e^+e^- \rightarrow \pi^+ + X$ the characteristics

$$\frac{q^2}{\sigma_{tot}}\cdot E\,\frac{d^3\sigma}{d^3p} = \frac{f_h(x)}{x^2} \quad \text{or} \quad \frac{1}{\sigma_{tot}}\cdot\frac{d\sigma}{dx} \simeq \pi\,\frac{f_h(x)}{x} \tag{3.46}$$

Experimentally this is not very well satisfied.

iv) EVMD (truncated) (38b) : The series of vector mesons might be truncated after some $n > N$. With the choice eq. (3.43) for f_n, Γ_n such an assumption leads to a scaling total cross section. The asymptotic hadron to μ-pair production ratio is linearly dependent on this cut-off parameter N

$$R_\infty \simeq \frac{9}{16} \cdot \frac{\Gamma_\rho \, \Gamma_{\rho \to e^+ e^-}}{\alpha^2 \cdot m_\rho^2} \cdot N \qquad (3.47)$$

and was estimated around $N > 20$.

In all these models the angular distribution is supposed to be isotropic and $< n >$ expected to grow linearly in q^2 . However, these assumptions are lacking specific models.

We indicate further possible developments in this direction (40) . So far the ℓ-particle decay width $\Gamma^{(\ell)}$ has not been specified. It could be fixed by models describing statistical decay of the resonance R_n . A Fermi model decay (see section 3.3.5) would suggest

$$\Gamma^{(\ell)}(q^2) = c_\ell (q^2)^{\ell-2} \qquad \Gamma_n^{(\ell)} \equiv \Gamma^{(\ell)}(m_n^2) \qquad (3.48)$$

$$\Gamma_{tot}(q^2) = \text{const.}(q^2)^{-\frac{4}{3}} \cdot \exp\{ \alpha \cdot q^{\frac{2}{3}} \}$$

(We are aware of the exponentially increasing total width and consider this model only as an example to illustrate the idea). The ℓ-particle and total cross section are

$$\sigma^{(\ell)} = 16\pi^2\alpha^2 \sum_n \frac{1}{f_n^2} \cdot \frac{m_n \cdot \Gamma_n^{(\ell)}}{(q^2-m_n^2) + m_n^2\,\Gamma_n^2} \qquad \sigma_{tot} = \sum_\ell \sigma^{(\ell)} \tag{3.49}$$

In addition, the choice of a particular decay model for $R_n \to$ hadrons permits specification of the inclusive distribution. The undetermined parameter, apart from the decay constants, is the coupling constant f_n .

Cross sections, described by the EVMD model, are expected to satisfy the sum rule

$$\int_{4m_\pi^2}^{s_{max}} ds \cdot s \;\{\, \sigma_h(s) - \sigma_{comp.}(s) \,\} = 0 \tag{3.50}$$

which permits a duality-like connection between the low energy resonance structure and the high energy behaviour of the total cross section; this connection is called "Sakurai-Duality". $\sigma_{comp.}$ is a comparison cross section, approaching σ_h for $q^2 \to \infty$, extrapolated down to low energies with appropriate threshold factors. The basic argument for such an approach is the superconvergence relation (deduced from unsubtracted dispersion relations) for the hadronic part of the vacuum polarization, whose imaginary part is related to the colliding beam cross section via : $\sigma_h = \frac{4\pi\alpha}{q^2}\ \mathrm{Im}\ \pi_h(q^2)$ [41] . One might go further and investigate the consequences of sum rules based on once-

or twice subtracted dispersion relations ; the result would be a negligible small constant (diffractive) component [42]. This result is in agreement with the finding of semi-local duality between the resonances ρ, ω and ϕ and the predictions of the quark-parton model. Stated differently : the point coupling of the virtual photon to the quarks averages on a semi-local level the resonance contributions of ρ, ω and ϕ [43].

For $q^2 \lesssim 9\ \mathrm{Gev}^2$ this model is quite attractive. The resonance widths grow with q^2 and the resonances become wider and wider interfering to a smooth curve ; however, the transition from the pronounced structure around $V_o \equiv (\rho, \omega, \phi)$ to the much smaller enhancements beyond V_o is quite fast. The new narrow resonances fit into this scheme if one assumes that they are composed of new types of quarks and thus form the beginning of a new series of meson resonances. However, if none are found at higher energies this scheme is in trouble.

3.3.4. Cuts [40]

Once one admits an infinite series of resonances with linear mass spectrum : $m_n^2 = m_o^2 \cdot (1 + a \cdot n)$, one may equally well assume that they are produced by the simultaneous emission of one or several other particles. Such a picture then assumes that the q^2-dependence of the matrix element, describing

<u>hadron production by an (off-shell) photon, is dominated by the onset of two-particle cuts and their threshold factors.</u> One might go further and investigate the influence of multiparticle cuts, however, we think the characteristic features will remain similar.

In a first attempt we neglect spin complications and only consider the threshold factor (for different masses m^2, M^2)

$$\sqrt{1 - 2\,\frac{M^2+m^2}{q^2} + \left(\frac{M^2-m^2}{q^2}\right)^2} \tag{3.51}$$

and the form-factor $F(m^2, M^2 ; q^2)$ (Fig. 44 .). The behaviour of the form-factor for large masses and large q^2 is problematic. In a first attempt one might set $F = 1$ and then finds an increasing hadron to μ-pair ratio : $R \propto q^2$. Let us suppose the form-factor scales for large variables as

$$F(m_n^2, m_n^2 ; q^2) \rightarrow f_o \cdot (m_n^2/ q^2)^{\beta} \tag{3.52}$$

For equal masses $m^2 = M^2 = m_n^2$, the model then predicts

$$R(q^2) = g \sum_{n=0}^{N(q^2)} f_o^2 \left(\frac{m_n^2}{q^2}\right)^{2\beta} \sqrt{1 - \frac{4m_n^2}{q^2}} \sim \left(\frac{gf_o}{4m_o^2 a}\right)\cdot q^2 \tag{3.53}$$

The unequal mass case, M^2 = fixed, $m^2 = m_n^2$

$$R(q^2) = g\cdot\sum_{n=0}^{N(q^2)} f_o^2\cdot\left(\frac{m_n^2}{q^2}\right)^{2\beta}\sqrt{1 - 2\,\frac{M^2+m_n^2}{q^2} + \left(\frac{M^2-m_n^2}{q^2}\right)^2} \propto q^2 \qquad (3.54)$$

does not essentially change the result. The mass M^2 does not appear in the asymptotic form.

Taking the opposite point of view that the mass spectrum is limited, $m_n^2 = M_o^2$, we find a cross section behaviour which is dependent on the form factor decrease :

$$R(q^2) \sim g\cdot f_o^2 \frac{(4M_o^2)^{2\beta+1}}{4m_o^2\cdot a(2\beta+1)}\cdot\left(\frac{1}{4q^2}\right)^{2\beta} \qquad (3.55)$$

The following global pictures present themselves :
In the q^2-region where new thresholds constantly open because of quasi two-particle intermediate states (with linear mass spectrum), $R(q^2)$ rises linearly in q^2 – independent of the form-factor decrease! R thus rises at most linearly with q^2. Once this region is passed and <u>no higher mass states</u> are created, its behaviour is dictated by the form-factor decrease (Fig. 45). If $R(q^2)$ should ever show a decrease at higher energies, this mechanism might be an explanation.

An alternative point of view is to assume that the <u>linear mass spectrum is never truncated</u>. If all form-factors (normalized to unity at $q^2 = 0$) are dominated in their decrease by that of $\pi\pi$-scattering

$$F_\rho(q^2) = \frac{m_\rho^2}{m_\rho^2 - q^2} \sim 1/q^2 \tag{3.56}$$

the ratio

$$R(q^2) \lesssim 1/q^2 \, . \tag{3.57}$$

This result means : the (finite) sum of threshold factors (below q^2) is asymptotically irrelevant. Such a result is in agreement with the finding of duality (as described in the preceeding section) between the resonance poles ρ, ω, ϕ,... and the asymptotic parton model (43). It shows that two-particle intermediate states indeed may be neglected.

Within the parton model framework this model is interpretable if a mass spectrum of point interacting partons is admitted. Could it form a bridge to the massive quark model (29d)?

$$R(q^2) = \sum_n P(m_n^2) \sqrt{1 - \frac{4m_n^2}{q^2}} \tag{3.58}$$

$P(m^2)$ is the probability to create partons of mass m_n.
The above linear mass spectrum is just one possibility. However, its origin has to be explained without resorting to composed objects, which would demand a form factor. One might give up this strict point of view and try fireball-like objects supposing that their interaction is described by field theory.

Other form-factor decreases could be possible such as

$$F(q^2) \sim \left(\frac{1}{q^2}\right)^{1 + J_n} \tag{3.59}$$

where $J_n = \alpha_o + \alpha' \cdot m_n^2$ à la Regge. If, in the model of eq. (3.53), the parton-hadron transition is supposed to scale, one expects

$$\frac{1}{\sigma}\cdot\frac{d\sigma}{dx} = \text{fct.}(x) \tag{3.60}$$

to scale. Such an assumption can be changed by choosing $x_n = 2E/\sqrt{m_n^2}$ as the relevant variable, for instance[40].

3.3.5. Fermi Statistical Model[44]

So far we have assumed that the photon transmits its energy to a small number of constituents only. One may take the opposite point of view and suppose that the photon interacts with many constituents. It converts, by some unknown process, its energy instanteneously into highly compressed and energetic "prematter" which reaches first statistical equilibrium and then explodes thus creating hadrons. This model is due to Fermi[45] (Fig. 46). The hadron creation process, supposed to be purely statistical, is described by the n-particle final state phase space. The matrix element for the transition $\gamma \to$ hadrons is then independent of any final state momentum p_i although there remains an unspecified q^2- dependence. It is the reason why statistical theories can never make predictions on the total cross section but only on normalized distributions. Using the input assumption $M_n = \kappa(q^2) \cdot (\sqrt{K})^n$, where K is a constant representing the coupling strength, the n-particle production cross section is given by the asymptotic form of the n-particle phase space (46)

$$\Gamma_n(q^2) = \frac{1}{n!} \int d^n\text{Phase space} \; |M_n|^2 \simeq \kappa^2 \frac{\pi}{2} \cdot \frac{(q^2 \frac{\pi}{2})^{n-2}}{(n-1)!(n-2)!} \cdot \frac{K^n}{n!} \qquad (3.61)$$

normalized to $\Gamma_{tot} = \sum_{n=2}^{max} \Gamma_n$ thus $\frac{\sigma_n}{\sigma_{tot}} = \frac{\Gamma_n}{\Gamma_{tot}}$ (3.62)

The normalized single particle distribution may then be worked out using

$$\frac{1}{\sigma} \left(\frac{d^3 \sigma}{d^3p/2E}\right) = K^2 \cdot \frac{d\,\Gamma(\{q-p\}^2)}{d\,K} \cdot \frac{1}{\Gamma_{tot}(q^2)} \qquad (3.63)$$

Its explicite form is (47)

$$\frac{1}{\sigma}\frac{d\sigma}{dx} = f \cdot \sqrt{x^2 - \frac{(2m)^2}{q^2}} \left(1 + \frac{4}{3}x\right) \cdot \exp\{3 \cdot f^{\frac{1}{2}}(g-1)\} \qquad (3.64)$$

$$f = f(q^2) = \frac{\pi K}{2} q^2$$

$$g = g(x;q^2) = \left(1 - x + \frac{m^2}{q^2}\right)^{\frac{1}{3}}$$

m is the mass of the created final state hadrons and $K \cong 15\ \mathrm{Gev}^{-2}$ was determined via the mean multiplicity : $< n > \cong \frac{2}{3} + (\frac{\pi K \cdot q^2}{2})^{1/3}$. Its shape is drawn in Fig. 47 . In the present energy range this simple phase space model represents quite well the normalized inclusive single particle distribution and permits comparison with the hadronic reactions $p\bar{p}$ (or $\bar{p}n$) $\rightarrow \pi$ + hadrons. For $q^2 \rightarrow \infty$, it increases (if $x \lesssim 0.35$) or decreases (if $x \gtrsim 0.4$) exponentially and does not have a finite scaling limit. The normalized inclusive distributions for different energies therefore cross around $x \sim 0.37$. The average multiplicity and average secondary energy grow with non-integer powers of q^2. The angular distribution is isotropic since there is no a priori distinguished direction (unless restrictions on the final state phase space are introduced through subsidiary conditions like eg. p_t-cut-off, etc.). Such attempts have to be taken as first approximations. They oversimplify the particle creation process by assuming that it is of purely statistical nature; quantum number restrictions have to be taken into account. There might be binary or triple interactions or even cluster formations (48).

Let us re-examine the matrix element. It contains as variables q and all final state momenta p_i and depends only on the relativistic invariants which can be formed by them

$$M(q; p_i) = M(q^2; p_i \cdot p_j ; p_i \cdot q) \qquad (3.65)$$

The use of the above formalism imposes the constraint of factorizability

$$M = \kappa(q^2) \prod_i f(q, p_i) \quad \text{or} \quad \kappa(q^2) \prod_{ij} f(q;p_i,p_j) \text{ etc.} \tag{3.66}$$

otherwise the phase space integrals become clumsy and difficult to solve in a closed form. Attempts in this direction have been tried. The resulting average multiplicites and average secondary momenta are listed in table V.

We point to an interesting feature (40). If the δ^4-function in the phase space integral

$$\Gamma_n(q^2) = \kappa(q^2)\cdot\frac{1}{n!}\int \frac{d^3p_1}{2E_1}\cdots\frac{d^3p_n}{2E_n} \cdot \delta^4(q-\Sigma p_i)\ |M_n(q,p_i)|^2 \tag{3.67}$$

is replaced by its Fourier representation

$$\delta^4(Q) = \frac{1}{(2\pi)^4}\int_{-\infty}^{+\infty} d^4x \quad e^{-iQ\cdot x} \tag{3.68}$$

and the matrix element is supposed to factorize as in eq. (3,66), Γ_n may be given the form

$$\Gamma_n = \kappa(q^2)\int_{-\infty}^{+\infty} \frac{d^4x}{(2\pi)^4} \quad e^{-iq\cdot x}\,\frac{1}{n!}\,\chi^n \tag{3.69}$$

with

$$\chi = \chi(x;q) \equiv \int_{-\infty}^{+\infty} d^4p\ \delta^+(p^2-m^2)\ f(p,q)\ e^{+ip\cdot x} \tag{3.70}$$

Summation over n exhibits :

$$\Gamma_F = \sum_{n=2}^{\infty} \Gamma_n = \kappa(q^2) \int_{-\infty}^{+\infty} \frac{d^4x}{(2\pi)^4} e^{-iqx} \{ e^{\chi} - 1 \} \qquad (3.71)$$

<u>The total cross section has an eikonal-like form</u>. In statistical bootstrap theories, describing chain-like decays (see below), the 1/n! -factor is absent and one obtains

$$\Gamma_{SB} = \kappa(q^2) \int_{-\infty}^{+\infty} \frac{d^4x}{(2\pi)^4} e^{-iq\cdot x} \{ \frac{\chi}{\chi-1} \} \qquad (3.72)$$

This form is reminiscent of the K-matrix formalism which permits a unitary representation of the T-matrix. If $f(q,p) = 1$, the "eikonal function" can easily be integrated and one notices the connection

$$\chi(x) = i\,(2\pi)^3\, \Delta^+(x) \qquad (3.73)$$

where $\Delta^{(+)}(x)$ is the Feynman propagator in x-space with the integration contour taken above the real axis; its explicit form is given in Ref. (49). This type of approach is interesting for several reasons :

i) If χ is supposed to have no dependence on the momentum q and $\kappa(q^2)$ = const. , it then permits link up with light-cone investigations. The current commutator, which determines the leading singularity, is identified with the eikonal function :

$$\langle 0 | [J_\mu(x), J^\mu(0)] | 0 \rangle = F(\chi(x)) \qquad (3.74)$$

Once $\chi(x)$ is specified the inclusive distribution is fixed.

ii) If there are several final state particles with different masses $m_1, m_2, \ldots$ the eikonal function $\chi(x)$ changes according to

$$\chi = \chi_1 + \chi_2 + \ldots \tag{3.75}$$

This result is found by summing :

$$\sum_{m=0}^{n} \frac{1}{m!} \frac{1}{(m-n)!} \chi_1^m \chi_2^{n-m} = \frac{1}{n!} (\chi_1 + \chi_2)^n \tag{3.76}$$

iii) Supposing there are binary interactions between the outgoing hadrons. This may be described as :

$$\Gamma_n = \kappa(q^2) \int \frac{d^4x}{(2\pi)^4} e^{-iq\cdot x} \binom{n}{2} \frac{1}{n!} (\sqrt{\chi})^n \tag{3.77}$$

$$\chi(q;x) = \int \frac{d^3p_1}{2E_1} \cdot \frac{d^3p_2}{2E_2} f(p_1,p_2) \cdot e^{ix\cdot(p_1+p_2)}$$

Summation over n gives

$$\Gamma_F = \frac{1}{2} \kappa(q^2) \int_{-\infty}^{+\infty} \frac{d^4x}{(2\pi)^4} e^{-iq\cdot x} \{ \chi \cdot e^{\sqrt{\chi}} \} \tag{3.78}$$

In the case of the statistical bootstrap model we find

$$\Gamma_{SB} = 2 \kappa(q^2) \int_{-\infty}^{+\infty} \frac{d^4x}{(2\pi)^4} e^{-iq\cdot x} \{ (\frac{\sqrt{\chi}}{1-\sqrt{\chi}})^2 + (\frac{\sqrt{\chi}}{1-\sqrt{\chi}})^3 \} \tag{3.79}$$

Such an approach relies on the specific form of the total cross section and the eikonal function $\chi\,(q;\,x)$ and thus leads to characteristic features in the inclusive distribution. The essential assumption is factorizability of the matrix element and of the phase space.

3.3.6. The Landau Statistical Model[46]

Fermi's model was further developed by Landau[50] who views the particle creation process in three stages :

a) pre-matter formation,

b) hydrodynamic-like expansion

c) hadron creation (Fig. 48)

The first step is as in Fermi's model. The photon converts its energy into statistical energy distributed over many constituents, viewed as pre-matter, in a finite volume of radius $r \sim 1/m_\pi$. The calculation concerning the expanding stage (supposing that the entropy is conserved) is made on the basis of the energy-momentum tensor of a perfect fluid

$$T^{\mu\nu} = (\varepsilon + p)\cdot u^\mu u^\nu - p\cdot g^{\mu\nu} \qquad (3.80)$$

$u^\mu(x)$ is the four-velocity field and $\varepsilon(x)$, $p(x)$ are the scalar distributions of energy and pressure. Pre-matter is treated as consisting of extremely relativistic constituents whose equation of state is $p = c_o^2 \cdot \varepsilon$. In the case of blackbody radiation - as supposed by Landau - $c_o^2 = 1/3$. The explicit calculation of the single particle inclusive distribution involves solution of eq.(3.80) and needs more explanation about the model, which we would like to skip and just state the result

$$\frac{1}{\sigma_{tot}} \cdot \frac{d\sigma}{dx} = \text{const.} \frac{x \cdot q^2}{\sqrt{1+\frac{4m_\pi^2}{x^2 \cdot q^2}}} \int d\eta \cdot dz \cdot \frac{dN}{d\eta} \cdot f(\eta,z) \qquad (3.81)$$

$dN/d\eta$ is the number of pre-matter constituents after expansion (where η is a parameter describing its evolution) which has to be convoluted with the energy distribution of a Bose-gas[51]

$$f(\eta,z) = \frac{\bar{E}}{\exp(\frac{\bar{E}}{kT_c}) - 1} \qquad \begin{aligned} \bar{E} &= E \cdot \cosh\eta - z \cdot p \cdot \sinh\eta \\ k \cdot T_c &= m_\pi \\ z &= \cos\theta \end{aligned} \qquad (3.82)$$

The numerically predicted curve is drawn in Fig. 49 . The following global features are inherent in this model : the inclusive distribution does not scale but rises with growing q^2 like

$$\frac{1}{\sigma} \cdot \frac{d\sigma}{dx} \sim (q^2)^{\frac{3}{8}} \qquad (3.83)$$

and concentrates more and more towards small x . It is insensitive to changes of the speed of sound c_o^2 of pre-matter. The average multiplicity and average energy of the final state particles increase as non-integer powers of q^2. The increase is dependent on the speed of sound c_o

$$\langle n\rangle \propto (\sqrt{q^2})^a \rightarrow (q^2)^{\frac{3}{8}} \qquad a \equiv \left(\frac{1}{1+c_o^2}\right)$$
$$\langle E\rangle \propto (\sqrt{q^2})^b \rightarrow (q^2)^{\frac{1}{8}} \qquad b \equiv \left(\frac{c_o^2}{1+c_o^2}\right) \tag{3.84}$$

The evaluated values are for $c_o^2 = 1/3$.

One of the essential ingredients in the Fermi- and Landau statistical models, applied to e^+e^- annihilation, is a constant initial state volume : V_o = const. On the basis of light cone arguments it could equally well be assumed to decrease as $V_o \sim E^{-3}$. Consequences then are : a constant multiplicity, a linear increase in $\sqrt{q^2}$ of the average momentum of the secondaries and an invariant inclusive distribution which is similar to a Bose-Einstein curve for small p , becoming broad as p increases [51h]. More recent developments take viscosity effects into account [52].

3.3.7. The Heisenberg Statistical Model

Heisenberg's statistical model (53), as another alternative in this class of approaches, has so far not been applied to e^+e^- annihilation. With its large energy the photon creates a shockwave in a system of sub-constituents (Fig. 50). It propagates radially outwards according to the equation

$$(\partial_\mu \cdot \partial^\mu + m^2) \phi + \alpha \phi^3 = 0 \qquad (3.85)$$

until the energy density is reduced to such an extent that hadronic particle creation becomes possible. m is the mass of the vacuum constituents and α a coupling parameter. The number of produced mesons was supposed to be proportional to the intensity of the wave front and therefore to the energy contained in the field. With suitable approximations (one dimensional treatment, etc.) this model was applied to hadronic reactions; we refer for details to Ref. (53c).

However, the exact solution to this equation is now known since the more general soliton wave equation

$$(\partial_\mu \cdot \partial^\mu + m^2) \phi + \alpha \phi^{2p+1} + \beta \phi^{4p+1} = 0 \qquad (3.86)$$

has recently been solved for $p \neq 0, -\frac{1}{2}, -1$ (54) :

$$\phi_{\pm}(z) = A_{\pm} \cdot e^{\mp iz} \cdot f(z) \qquad z = k_o \cdot x_o - \vec{k} \cdot \vec{x} \tag{3.87}$$

$$f(z) = \left[\left\{ 1 - \alpha \frac{(A_{\pm})^{2p} \cdot e^{\mp i2pz}}{4m^2(p+1)} \right\}^2 - \beta \frac{(A_{\pm})^{4p} \cdot e^{\pm i4pz}}{4m^2(2p+1)} \right]^{-\frac{p}{2}}$$

It permits more rigorous treatment [40].

3.3.8. The Thermodynamic Model [44,46,55]

This model (also called statistical bootstrap or Hagedorn model) is distinct from the other statistical model approaches in its decay dynamics. Instead of a single center decay, the photon is supposed to convert to a fireball which gradually desintegrates into hadrons and/or fireballs of lower masses (Fig. 51). Fireballs may be defined as privileged regions in the multiparticle phase space of hadrons or constituents. Their origin is either due to usual dynamics (singularity in the S-matrix, etc.) or due to some cooperative multiparticle effects. These are certainly of dynamical nature, but are described directly through the phase space modifications.

Description of statistical chain decays as drawn in Fig. 52 are similar to the phase space expression Γ_n (eq. 3.61) of the Fermi model, 1/n! however, is missing since the n finalstate hadrons may be replaced in their relative positions along the chain. Immediate consequences are : a constant average energy per created hadron and a multiplicity increase linear in E_{CM} :

$$\langle n\rangle \sim \sqrt{q^2} \qquad \langle p\rangle \sim \text{const.} \qquad [\,\sigma_n = d_n\cdot(q^2)^{n-2}\,] \tag{3.88}$$

The normalized inclusive distribution[56a]

$$\frac{1}{\sigma}\frac{d\sigma}{dx} = f(kT_o)\cdot(\sqrt{q^2})^3\cdot\{x^2-\frac{4m_\pi^2}{q^2}\}\cdot\exp(-\frac{\sqrt{q^2}}{2kT_o}\cdot x) \tag{3.89}$$

does not scale but rather increases like $\sim (q^2)^{3/2}$ and the x-dependence shifts to smaller x-values. Such features were also found in the Landau statistical model. Needless to say the angular distribution is isotropic. There are three interesting features to be noticed :

First. The model predicts that the distribution

$$\frac{1}{\sqrt{q^2}}\cdot\frac{1}{\sigma_{tot}}\,E\cdot\frac{d\sigma}{p^2dp} \simeq \frac{2}{q^2}\cdot\frac{1}{p}\cdot\frac{1}{\sigma_{tot}}\cdot\frac{d\sigma}{dx} \overset{!}{\equiv} \text{Fct.}(E) \tag{3.90}$$

is only dependent on the energy of the inclusive particle E and does not change with increasing initial energy $E_{CM} = \sqrt{q^2}$. In the present energy range, the experimental results do not give support for this hypothesis.

Second. In the inclusive distribution (3.89) the temperature (or mass) of the fireball created by the photon appears explicitly. In Fig. 53 we have drawn for $k\cdot T_o = 193$ Mev the distributions which for smaller values of the parameter

$k \cdot T_o$ considerably rise[56]. For a simple exponential ansatz, the data suggest such a value.
Note that in all statistical models the initial temperature (which sometimes is hidden in the formalism) is responsible for scale-breaking since it has the dimension of a mass.

Third. An inherent property of the thermodynamic as well as Fermi model (inv. phase space form) is an exponentially decreasing ratio

$$\frac{\sigma_n}{\sigma_{tot}} = \begin{cases} c_n (q^2)^{n-\frac{8}{3}} e^{-\alpha\sqrt{q^2}} & \text{(Fermi)} \\ d_n (q^2)^{n-\frac{5}{4}} e^{-\beta\sqrt{q^2}} & \text{(Thermodynamic)} \end{cases} \tag{3.91}$$

with increasing CM-energy.

We briefly indicate generalizations of this model and its connection with thermodynamics[44,55]. The n-particle phase space may be split into two or more sub-spaces

$$\Gamma_n(q^2) = \int \prod_i^n \frac{d^3p}{2E_i} \delta^4(q-\sum^n p_i) = \int \Gamma_1(p^2)\cdot\Gamma_{n-1}(k^2)\cdot\delta^4(p+k-q)\, d^4p\cdot d^4k \tag{3.92}$$

Single-particle phase space is then $\Gamma_1(p^2) \equiv \delta^{(+)}(p^2 - m^2)$.
Let us consider the decay : fireball → fireball + particle. Such a process may continue until energy is left for one final hadron only. In formal language it is described by the integral equation

$$\Gamma(q^2) = \Gamma_1(q^2) + \lambda \int \Gamma_1(p)\cdot\Gamma(k^2)\, \delta^4(p+k-q)\, d^4p\cdot d^4k \tag{3.93}$$

which may be solved by taking the 4-dimensional Laplace transform

$$Z(\beta,\lambda) \equiv \int d^4q \cdot e^{-q^\mu \cdot \beta_\mu} \cdot \Gamma(q^2) \qquad (3.94)$$

Note the connection of this step with the Fourier transformation proposed in section 3.3.5. The form of this grand canonical partition function $Z(\beta, N)$ determines the essential characteristics of the model once one identifies

$$\bar{E} = -\frac{d}{d\beta} \ln Z \equiv \sqrt{q^2} \quad \Rightarrow \quad \begin{array}{l} \langle n \rangle = 1 + \lambda \dfrac{d}{d\lambda} Z \\ \langle p \rangle \simeq \dfrac{\sqrt{q^2}}{\langle n \rangle} \end{array} \qquad (3.95)$$

and permits definition of the fireball temperature. The essential characteristics remain unchanged in this bootstrap formulation.

Within the same spirit of the above thermodynamic model, there exist some models which assume multipion production as the result of a vector meson cascade involving the ρ and ω mesons (57). Such picture supposes the production of pions to take place via a kind of "bremsstrahlung" process, however, with the restriction that the spin-1 system decays into a pion plus another spin-1 system. The asymptotic behaviour of the total hadronic annihilation cross section is closely related to the asymptotic off-shell formfactor and can be chosen accordingly : $\sigma(e^+e^- \rightarrow h) \sim s^{-3}$, exponentially increasing, $s^{\alpha(\lambda) - 5/2}$. The trajectory $\alpha(\lambda)$ in the third case is determined by an eigenvalue equation, starts from $\alpha(0) = 0$ and increases monotonically the average pion multiplicity then grows logarithmically. The spin restriction on the "fireball" in this model leads to observable differences with respect to the bootstrap models above.

In the Bj-limit a linear combination of the structure functions (of the single particle inclusive distribution) : $F = W_1 - \frac{\omega}{6} W_2$ obeys Feynman scaling for $\omega > 0$ fixed : $F(s, \omega) = s^{\alpha} F(\omega)$, however, the scaling function $F(\omega)$ exhibits a singularity at $\omega = 0$; thus the limit $s \to \infty$, $\omega \to 0$ is in general non-uniform. A nice feature: although there is no difference in the charged and neutral multiplicities the major proportion of the CM-energy is carried away by neutral pions.

3.3.9. The Uncorrelated Jet Model (58)

An attempt, different in its mathematical form from the above statistical bootstrap model, however, similar in spirit, proposes that particles are statistically created in eventwise jets, the jet axis being randomly distributed between 0 and 2π (Fig. 54). The underlying physical picture of this uncorrelated jet model assumes the production of a pair of fireballs which subsequently decay into hadrons.

The explicit form of the single particle inclusive distribution

$$\frac{1}{\sigma}\left(\frac{d^3\sigma}{d^3\frac{p}{2E}}\right) = \frac{\int d^2e\cdot K\cdot e^{-\lambda|\vec{p}x\vec{e}|}\ \Gamma(\vec{e},q-p)}{\int d^2e\cdot\Gamma(\vec{e},q)} \qquad (3.96)$$

$$\Gamma(\vec{e},q) = \sum_{n=2}^{\infty}\frac{\kappa^n}{n!}\int\prod_i^n\frac{d^3p_i}{2E_i}\cdot e^{-\lambda|\vec{p}_i x\vec{e}|}\ \delta^4(q-\Sigma\, p_i) \qquad (3.97)$$

($\vec{e}$ is a unit vector specifying the jet axis of each event)

exhibits its asymptotic scaling behaviour.

$$\frac{1}{\sigma}\frac{d\sigma}{dx} \simeq 2\bar{K}\frac{(1-x)^{\bar{K}-1}}{(x^2-\frac{4m^2}{q^2})^{\frac{1}{2}}}\ r^2\int_0^{\frac{\pi}{2}} d\theta\cdot\sin\theta\cdot\ e^{-r\cdot\sin\theta} \qquad (3.98)$$

$$r = \frac{\lambda}{2}\sqrt{q^2}\cdot(x^2-\frac{4m^2}{q^2})^{\frac{1}{2}}$$

It is expected to set in first far away from 0 and 1

$$\frac{1}{\sigma}\frac{d\sigma}{dx} \simeq 2\bar{K}\frac{(1-x)^{\bar{K}-1}}{x} \simeq \frac{6}{x}(1-x)^2 \qquad (0 \ll x \ll 1) \qquad (3.99)$$

The parameter λ in the exponential p_t-cut-off was determined by application of this model to high energy hadron-hadron scattering. From 90° pion spectra of $pp \to \pi^o + X$ one has $\lambda = 6.2\ \text{Gev}^{-1}$. The "decay-constant" K was fixed by the mean multiplicity suggesting $\bar{K} = \frac{\pi K}{\lambda^2} \simeq 3$. Representative indusive distributions are drawn in Fig. 55 . The mean multiplicity rises logarithmically whereas the mean momentum depends linearly on the CM-energy

$$\langle n\rangle \to \bar{K}\cdot\ln q^2$$
$$\langle p\rangle \to \frac{\sqrt{q^2}}{\ln q^2} \qquad (3.100)$$

Another distinct feature of this model is the asymptotic power decrease of the ratio : n-particle cross section to total cross section

$$\frac{\sigma_n}{\sigma_{tot}} = c_n \cdot (q^2)^{-\bar{K}} \left[\ln \frac{q^2}{4m^2} \right]^{n-1} \approx \left(\frac{1}{q^2} \right)^3 \tag{3.101}$$

It has to be compared with an exponential fall-off in the Fermi- and Thermodynamic model. Using the mean multiplicity it may be given the form

$$\frac{\sigma_n}{\sigma_{tot}} \approx \frac{\langle n\rangle^{n-1}}{(n-1)!} e^{-\langle n\rangle} \tag{3.102}$$

This model assumes an exponential p_t-cut-off. What about a power decrease as in large p_t hadronic reactions?

3.3.10. Cascade Decay Model (59)

The cascade decay model may be considered as another specific realization of the thermodynamic model. The photon here is supposed to convert its energy into a fireball of mass $\sqrt{q^2}$ - at rest - which subsequently decays into n sub-fireballs of mass $\sqrt{q^2}/n$ - again at rest. The probability for such a process is denoted as P_n . Each of these fireballs decays into other fireballs with the same probability. This process continues k-times; after the k-th step all fireballs decay into hadrons. The function $k = k(\sqrt{q^2}) = a \cdot \ln(q^2/m^2)$ was postulated and is lacking physical interpretation (Fig. 56).

This process is described by the mathematical theory of branching processes. It would be going to far to explain its content here ; we therefore concentrate on its predictions and characteristic features. The inclusive single particle distribution is of Gaussian shape in the variable $y = \ell n\, \omega$ $(\omega \equiv x^{-1})$

$$h(y,q^2) \equiv \frac{1}{<n>}\cdot\frac{1}{\sigma_h}\frac{d\sigma}{dy} \simeq \frac{1}{\sqrt{2\pi}\,\sigma_o}\cdot\exp\{-\frac{(y-\bar{y})^2}{2\sigma_o^2}\} \qquad (3.103)$$

(note this statement is P_n-independent) and satisfies the sum rule indicating scaling violation

$$\int_0^1 \frac{1}{\sigma_h}\cdot\frac{d\sigma}{dx}\cdot x^n\cdot dx = A(n)\cdot\left(\frac{M^2}{q^2}\right)^{\alpha(n)} \qquad (3.104)$$

The functions $A(n)$, $\alpha(n)$ are connected with the inclusive distribution of the decay : fireball → hadrons, and the generating function $Z(n) = \sum_1^\infty k^{1-n}\cdot P_k$. M^2 is an unknown mass-scale and $\alpha(n)$ an increasing function for increasing argument $n = 1,2,3,\ldots$ (59c). Within the framework of scale invariant field theories $\alpha(n)$ is interpretable and related to the anomalous dimensions and the short distance behaviour of this model.

Comparison of the form (3.103) with the experiment at E = 3.0 Gev and 4.8 Gev is excellent (Fig. 57). $\alpha(n)$ may be phenomenologically determined via the sum rule (3.104) .

The form $\alpha(n) = 1 - \frac{\lambda}{\lambda+n}$ was proposed, where λ remains an adjustable parameter. The connection of the inclusive distribution with $<n>$ together with the sum rule (3.104) exhibits the power behaviour of the average multiplicity and average momentum as :

$$\begin{aligned} <n> &\sim (q^2)^{-\alpha(0)} \\ <p> &\sim (q^2)^{1+\alpha(0)} \end{aligned} \tag{3.105}$$

Note that the multiplicity is constant for the above choice of $\alpha(n)$. There are three further characteristics to be noticed in this model.

First. The normalized n-particle production cross section obeys KNO-scaling[59f]

$$\frac{\sigma_n}{\sigma_{tot}} \simeq \frac{1}{<n>} \Phi\left(\frac{n}{<n>}\right) \tag{3.106}$$

Second. All n-particle cross sections have a universal asymptotic decrease like

$$\lim_{q^2\to\infty} \frac{\sigma_n}{\sigma_2} = \xi(n) \tag{3.107}$$

Third. The two particle production cross section decreases according to

$$\frac{\sigma_2}{\sigma_{tot}} = \left(\frac{q^2}{M^2}\right)^{a\cdot\ln P_1} \qquad (0 \le P_1 \le 1) \tag{3.108}$$

Thus, the form-factor decreases like a power in q^2 .

How distinct is this model from the statistical bootstrap model, apart from the above properties ? A model calculation predicts (59b)

$$P_n \propto \frac{u^n}{\Gamma^2\{n(\alpha+1)\}} \tag{3.109}$$

to decrease rapidly for large n . It has its maximum value at $n_{Max} \sim \frac{u^{\frac{\alpha+1}{2}}}{\alpha+1}$ - a value which may be adjusted.

The statistical bootstrap model on the other hand predicts $n_{Max} = 2$ with $P_2 \sim 0.7$; there is one heavy fireball and all successive ones in the chain have smaller masses. Energy is equally distributed among the fireballs in the cascade model and they are supposed to move slowly in the CM-system.

3.3.11. New Interactions

The non-scaling behaviour of the e^+e^- total cross section data for $q^2 \gtrsim 9$ Gev^2 initiated a few suggestions in unconventional directions.

a) Leptons with hadron behaviour (60)

One suggestion supposes that the interacting leptons show characteristics which so far were believed to exist only in the hadronic world (extended or composite structure, Regge asymptotics, Pomeron exchange, Diffraction, etc.).

Thus at high energies Pomeron-exchange would considerably influence the behaviour of the annihilation cross section (Fig. 58). Going to the extreme by postulating that it should dominate, scaling should never set in.

On the basis of such arguments, the asymptotic behaviour was found to be well parametrized by

$$\sigma_h(q^2) \simeq 2\pi G_F \tag{3.110}$$

where $G_F \cong 10^{-5}/m_p^2 \approx 4$ nb is Fermi's constant for weak interactions. For the inclusive single particle distribution the ansatz, taken from the purely hadronic reaction $pp \to \pi + X$,

$$E\cdot\frac{d^3\sigma}{d^3p} = A\cdot\exp\{-a\cdot p_t - b\cdot x_{//}^2\} \tag{3.111}$$

was proposed where p_t and $p_{//}$ are the momentum components of the inclusive particle (perpendicular and parallel to the direction of the initial e^+e^- pair) and $x_{//} \equiv \frac{2p_{//}}{\sqrt{q^2}}$.
The inclusive angular distribution shows as characteristics

$$\frac{d\sigma}{d\Omega}(\theta=\tfrac{\pi}{2}) \sim \text{const.}\ , \qquad \frac{d\sigma}{d\Omega}(\theta=0,\pi) \sim q^2 \tag{3.112}$$

Consequently, one expects a strong peaking of the angular distribution in the forward and backward directions which is experimentally disproved.

The mean multiplicity grows logarithmically and in consequence the mean momentum grows linearly in q^2 . The inclusive distribution does not scale. As characteristic features we mention

$$\frac{d\sigma}{dx_{\parallel}} = \begin{cases} \text{const.} \ \sqrt{q^2}\, e^{-b\cdot x_{\parallel}^2} & \text{for} \ x_{\parallel} = \frac{2p_{\parallel}}{\sqrt{q^2}} \ \text{small} \\ \text{const.} \ \frac{e^{-b\cdot x_{\parallel}^2}}{|x_{\parallel}|} & \text{for} \ x_{\parallel} \ \text{large} \end{cases} \tag{3.113}$$

with an exponentially falling large p_t-dependence.

Further consequences of such new interactions might be(64) :

1) $\sigma(e^+e^-)$ and $\sigma(e^-e^-)$ might become flat and tend towards the same asymptotic value.

2) One expects <u>no</u> leptons in the final state of $e^-e^- \rightarrow$ hadrons.

3) Elastic e^+e^- and e^-e^- scattering should show effects which depart from usual QED predictions. Tests sofar were up to $r \sim 10^{-15}$cm .

4) $g_e - 2$ should show deviations form QED.

b) Direct Lepton-Hadron Coupling(61)

Can leptons pull hadronic constituents out of the vacuum by direct interaction without the intermediary of a photon or any other exchanged object ? The form of a 4-fermion interaction has been suggested whose effective Lagrangian is

$$\mathcal{L}_I = \sum_i g_i (u_{\bar{e}} \Gamma_i u_e) \cdot (u_{\bar{q}} \Gamma_i u_q) \tag{3.114}$$

u_e and u_q are the Dirac spinors of the leptons and hadronic constituents (quarks, partons, etc.) and the Γ_i are any independent matrix of the Dirac algebra. Such a Lagrangian is non-renormalizable since $[g^2] = 1/M^2$. Ways out would here be as in weak interactions. The cross section behaves as

$$\sigma_h = \frac{4\pi\alpha^2}{3q^2} (\sum_i Q_i^2) \cdot \{ 1 + 2 \cdot \delta_1 \cdot q^2 + \delta_2 \cdot q^4 \} \tag{3.115}$$

and the effective coupling constant is estimated in the range

$$g_i \approx 3 \cdot 10^{-3} \ (\mathrm{Gev})^{-2} \tag{3.116}$$

These schemes of <u>direct coupling of leptons to $q\bar{q}$ have the disadvantage that the final state distribution of hadrons in the direct process would be expected to be very similar to the one-photon process</u>.

<u>c) Lepto -hadron</u>(62)

The afore mentioned 4-fermion interaction might be considered as the limiting case of a peripheral reaction $e^+e^- \to q\bar{q}$. The leptons hit on hadronic constituents by exchange of a lepto -hadron $X(\ell, h)$. It has to carry hadronic (h) as well as leptonic (ℓ) quantum numbers and must have spin 0 or 1 (Fig. 59).

The Lagrangian for such an interaction is

$$L_I = f\cdot(\bar{u}_e\gamma^\mu u_q)\cdot X_\mu + hc. \tag{3.117}$$

The total hadronic annihilation cross section, which has the same form as in eq.(3.115), strongly violates scaling due to the X-meson term and its interference with the scaling parton-model contribution. To obtain a sizeable contribution in e^+e^- annihilation the coupling constant for such a new interaction was estimated in the order of

$$\frac{f^2}{4\pi}\left(\frac{1}{m_X^2}\right) \simeq (\varepsilon\cdot\alpha)\ \mathrm{Gev}^{-2} \qquad \left(\varepsilon \simeq 0.02 - 0.1\ ,\quad \alpha \simeq \frac{e^2}{4\pi}\right) \tag{3.118}$$

and the application of these ideas to $\pi^o, \eta \to e^+e^-$ decays demands $m_X \gtrsim$ Gev. The large mass m_X justifies the neglecting of any crossed-channel variable dependence - but then we are back at the 4-fermion interaction which has about the same characteristics. Note, that the above value for the coupling constant f is considerably above the estimated one for the neutral current coupling : $\frac{g^2}{4\pi}\left(\frac{1}{M_{Z^o}}\right) \cong 3\cdot 10^{-7}$ Gev^{-2} !

If the inclusive distribution $e^+e^- \to \pi + X$ is dominated by the X-meson exchange, the angular distribution

$$\frac{d\sigma}{d\cos\theta} = \frac{1}{32\pi} \sum_i f_i(q^2)\cdot A_i\,(q^2, \cos\theta) \tag{3.119}$$

with $A_i(q^2, \cos\theta) \propto q^2\cdot\frac{g_i^2}{2}\{4 + (1 + \cos\theta)^2\}$

rises strongest at $\theta = 0, \pi$ (forward and backward direction) and much less at $\theta = \pi/2$. However, in the present energy range the interference terms cannot be neglected and the distribution is much more complicated. The inclusive distribution

$$q^2 \cdot \frac{d\sigma}{dx} = \frac{1}{32\pi} \sum_i D_i(x) \cdot f_i(q^2)$$
$$f_i(q^2) \equiv (a \cdot q^4 + bq^2 + c + d \cdot \frac{1}{q^2})_i \qquad (3.120)$$

does not scale but shows an increase, as in the total cross section. Besides the characteristics just sketched, hadron creation is expected to show features similar to the parton model $\langle n \rangle \sim \ell n\, q^2$ etc. Jet-like structure is likely to exist because of the directly produced $q\bar{q}$-pair. Multiperipheral characteristics also could exist in this model (Fig. 60).

Do these new interactions lead to any consequences in deep-inelastic electron-proton scattering ? Indeed one expects a measurable difference between e^+p and e^-p in both elastic and deep-inelastic scattering. If the structure functions due to single photon exchange and new interactions both are supposed to scale, the inelastic cross section will then be multiplied by a factor $\phi(q^2) = [1 + \varepsilon_1 q^2 + \varepsilon_2 q^4]$ where ε_1 and ε_2 are both of order α . In the present energy range one would therefore expect scaling violation of the differential cross section not larger than 10% (61,62).

d) Direct Channel "Objects"[63]

One might assume that the photon is not the only intermediate direct channel object to carry away the leptonic energy after their annihilation ; the e^+e^- annihilation process might also proceed via a neutral intermediate boson Z^o or a (Higgs) scalar meson ϕ (Fig. 61). However, these contributions are expected to be much smaller than the one photon exchange, except near their masses. Assymptotically one gets scaling in such a picture unless two ϕ-mesons (with low masses) accompanied by a vector meson are created that can generate a constant or slowly rising total cross section. Another consequence of direct channel scalar (besides one-photon) exchange is a time dependence of the total cross section[63a], due to changing polarization of the beam, according to

$$\sigma_h^P = \sigma_\gamma + (1 \pm |P|^2) \sigma_\gamma \qquad (3.121)$$

P is the time dependent polarization coefficient of the beam and σ_γ respectively, σ_ϕ represent the cross sections for single photon and scalar exchange. Experimentally no such effect could be found at SPEAR I since the polarization time constant was too large ; at SPEAR II it will be as short as 10 Min.[64a].

3.3.12. Models : General Features

In the preceeding subsections, we have presented the modelling approaches in e^+e^- annihilation in their established languages. We now would like to extract their essential structure and compare their characteristics. The above presented models may be classified in three types :

i) The off-shell photon (or lepton pair) creates a constituent pair ($q\bar{q}$) which subsequently creates hadrons. The first process is described by QED whereas the constituent - hadron transformation is parametrized by a phenomenological function $D_i(x)$ which only depends on the scaling variable $x \equiv \frac{2E}{\sqrt{q^2}}$.

ii) The off-shell photon creates many constituents (at once) which cooperatively interact and form a short-lived compound. Its dynamical behaviour is described by various models within the framework of statistical mechanics ; it is treated as a gas or fluid. The constituent-hadron transformation is unspecified - the essential input is a hadronic momentum distribution.

iii) The off-shell photon creates an infinite series of vector mesons (viewed as $q\bar{q}$-bound states) which decay into hadrons. The decay distribution is supposed to scale.

All models may be characterised by the following diagram

Photon $\overset{①}{\rightarrow}$ Constituents $\overset{②}{\rightarrow}$ Hadrons

Parton models do specify the transition ①. It is responsible for the q^2-dependence of the cross section, in particular, its scaling behaviour, and the inclusive angular distribution. The second transition is described by a phenomenological function which only depends on the scaling variable x. Thus : ① gives the q^2-dependence whereas ② the momentum distribution of the inclusive hadron. All scaling violation is blamed on the transition ① by supposing that either partons are not point-like or there is some new form of interaction.

In the EVMD-model the characteristics are similar. Transition ① : $\gamma \rightarrow R_n$ is described by a coupling constant f_n, which may or may not depend on q^2. The propagators essentially dominate the q^2-dependence of the cross section. Transition ②, the decay $R_n \rightarrow$ hadrons, is again supposed to scale (in some models); it depends only on the scaling variable x. This model has a large amount of flexibility because of its unspecified decay distribution and γ-R_n coupling constant respectively decay width $\Gamma(R_n \rightarrow e^+e^-)$.

Statistical models do not specify transition ① instead. A statistical ensemble of constituents with energy E_{CM} is assumed which evolves according to specific pictures : hydrodynamic expansion, fireball decays, etc. The statistical energy is thus distributed according to these pictures. The essential input is transition ② in the form of a momentum distribution of the final state hadrons. In the Fermi model it is purely statistical, whereas the Landau model assumes a Bose-distribution. The cascade model does not specify such distribution apart from a supposed scaling behaviour.

One of the common characteristics of all models is the assumption of hadron creation through the intermediary of constituents. This fact is closely linked with Dirac's picture of the physical vacuum. Recently, such fundamental assumption was questioned and different models for the vacuum were proposed : anti-particles in the sea might themselves interact or the sea might in fact be treated as a multi-component quantum liquid[(27b)].

In tables VI,VII,VIII we have listed the properties of the existing models which we will briefly compare. The inclusive angular distribution of the parton model is determined by the spin and magnetic momentum of the vacuum constituents. Models which suppose that many constituents are created at the same time do not give preference to any space direction.

(The uncorrelated jet model describes hadron production).
A scaling constituent → hadron transition function, sufficiently singular at small x, leads to a logarithmic (or stronger) increasing mean multiplicity and consequently to a mean momentum which is roughly power behaved. The former property is due to the connection of the mean multiplicity and the inclusive distribution via energy-conservation sum rule. Most thermodynamic models predict a non-integer power increase of the mean multiplicity - excluded are those which give asymptotic scaling. In parton-like model all scaling violations in the inclusive (and total) cross sections are due to scaling violating $q\bar{q}$-creation. $q\bar{q}$ → hadrons are always supposed to scale. The ratio σ_n/σ_{tot} is not specified in the parton model. Fermi and Thermodynamic model predict an exponential decrease for asymptotic $E_{CM} = \sqrt{q^2}$ and a power increase like $(q^2)^n$ at moderate energies ; n is the number of produced secondaries. The uncorrelated jet model grows as $(\ln q^2)^n$, whereas the cascade decay model is practically q^2-independent. The Heisenberg and Landau models do not specify σ_n/σ_{tot} .

4. THEORIES ON THE NEW RESONANCES

In this chapter we sketch the theoretical explanations for the new resonances. There are two unusual features to cope with : the unexpected high masses M_{ψ} and the extremely small total widths Γ_{ψ} which are roughly a factor 10^{-3} smaller than the ones with which one is familiar from the hadron decay channels :

$$\Gamma_{\psi} \sim 10^{-3} \cdot \Gamma_{V_0} \quad , \quad \Gamma_{V_0} \sim 1\text{-}100 \text{ Mev} \quad , \quad V_0 \equiv (\rho,\omega,\phi) \qquad (4.1)$$

The offered explanations may be split into three groups

a) hadronic - higher symmetry

b) hadronic - dynamical

c) non-hadronic

The first group contains all higher symmetry attempts ; new quantum numbers and/or selection rules are introduced which disfavour the hadronic decays. Models in the second group assume they are of hadronic nature but their small widths are due to dynamical reasons, whereas attempts in the third group assume that these resonances are of non-hadronic nature : weak interactions, new interactions, etc.

4.1. Hadrons : Higher Symmetry Schemes

The higher symmetry attempts lead to a wealth of predictive information from which we try to extract the essential points. These models assume that a bound state or resonance of quarks is formed, similar to ρ, ω, ϕ,..., etc. The decay dynamics is determined by symmetry arguments and selection rules in the space of quantum numbers.

One can essentially distinguish three lines of reasoning :

i)	four quarks - one of them with charm	:	SU(4)
ii)	three colored quarks	:	SU(3) x SU(3)
iii)	other new quantum number schemes		

Much effort has been invested in systems like i) and ii) which, however, have to be supplemented by dynamical schemes in order to explain the small ψ-widths. They are extremely useful in classifying a wealth of anticipated new particles and in correlating the various decay channels and masses.

4.1.1. Charm[65,66,67]

The attempt to unify weak and electromagnetic interactions in a renormalizable theory suggested the existence of a new quantum number "charm" (conserved by strong and electromagnetic interactions) to exclude strangeness changing (hadronic) weak neutral currents $\Delta S \neq 0$. Experimental motivations are[68] α) $\Gamma(K^+ \to \pi^+ e^+ e^-)$ / $\Gamma(K^+ \to \pi^o e\nu) \sim 10^{-5}$

β) $\Gamma(K_L \to \mu^+\mu^-)$ << $\Gamma(K_L \to \gamma\gamma)$

γ) Order of magnitude of $\Delta m = m_{K_L} - m_{K_S}$

It was conjectured that there exists a fourth quark ($s = \frac{1}{2}$, $B = 1/3$, $S = 0$, $I = 0$, $Q = 2/3$, $C = +1$ and mass $\lesssim$ few Gev) carrying this quantum number, whereas all other quarks are supposed to have charm zero. The Gell-Mann-Nishijima formula is then enlarged to

$$Q = I_3 + \tfrac{1}{2}\,(S + B + C) \tag{4.2}$$

(see Table IX).
The basic symmetry in strong interactions then becomes SU_4 instead of SU_3 . As a consequence, one expected for some time - before the discovery of the narrow resonances - that there should be effects of this kind due to charm[69].

The hypothesis that the narrow resonances are bound states of charmed quarks, whose masses are considerably larger than the masses of the conventional quarks, is quite attractive and has lead to a number of interpretations and decisive predictions we will present here [70,71,72]
The ideas presented in this section are closely linked with the dynamical investigations on the characteristics of a $c\bar{c}$-bound state which (for reasons of organization of this paper) are discussed in section 4.2.3 . The two sections however are complementary and therefore should be read together .

1) One expects the $\psi(3.1)$ resonance to be a $c\bar{c}$-bound state (vector meson) $\psi_c \equiv (c\bar{c})$ and the $\psi'(3.7)$-resonance to be its radial excitation since there is no ready explanation in SU(4) for the appearance of two states. Their narrow widths can be qualitatively understood if they lie below the threshold for production of a pair of charmed hadrons. A simple estimate of the widths due to non-charmed hadron production gives far too big values which, by some mechanism, have to be reduced by a factor $\sim 10^{-3}$. Zweig's selection rule (section 4.2.4) or "Charmonium" (section 4.2.3) are possible dynamical explanations. Note ψ_c states could in principle also be mixed with $\phi = \lambda\bar{\lambda}$ and $\omega = \frac{1}{\sqrt{2}} p\bar{p} - n\bar{n})$ etc., but are supposed to be pure $c\bar{c}$.

2) A number of further narrow resonances (with zero charm ≡ hidden charm) are anticipated below $c\bar{c}$-threshold due to orbital and radial excitations (Table X)(80e). Those with quantum numbers different from $J^{PC} = 1^{--}$ cannot be directly produced in e^+e^- annihilation via single photon exchange (due to charge conjugation and parity constraints) but in hadronic ($p\bar{p}$, etc.) reactions instead. (None have been found so far). The electromagnetic transition among the states in Table X should result in numerous monochromatic photons with various energies in the range of several tens to several hundreds of kev (Figs. 62). <u>The detection of such photons will be crucial in verifying the charm scheme!</u> (None have been seen so far).

3) In particular a pseudoscalar ($J^{PC} = 0^{-+}$) $c\bar{c}$-bound state ($\equiv \eta_c$) is expected with mass $m_{\eta_c} \cong 3$ Gev which decays into two photons (73). There are three questions:

a) Mass of η_c :

Its mass is expected not too far from ψ_c (possibly below). Higher symmetry arguments suggest $m_{\eta_c} = 3.122$ Gev, 3.066 (73a), $m_{\eta_c} = m_{\eta_c} - 0(10\text{Mev})$ (106), 80 Mev $\lesssim m_{\psi_c} - m_{\eta_c} \lesssim 400$ Mev (107a).

b) Production of η_c (73d):

Production is expected in

- $(\psi_c', \psi_c) \to \eta_c + \gamma$
- hadronic reactions : $p\bar{p}$, $\Lambda\bar{\Lambda}$ best
  : $K^{\pm}\pi^{\pm}K_S^o$, $\pi^{\pm} - A_2^{\mp}$ also possible.
- Primakoff photoproduction
- $e^+e^- \to e^+e^- + \eta_c$

Based on $\Gamma(\psi_c \to \eta_c + \gamma) = \tilde{\Gamma} \cdot (\frac{p\gamma}{m_{\psi_c}})^3$ the decay width seems to be strongly η_c-mass dependent, but is expected in the kev-region. Typical values are represented in Table XII; for comparison we have also listed the known values of usual hadronic decays in Table XIII. The decay $\psi_c \to \eta_c + \gamma$ might be suppressed; in

$$\Gamma(\psi_c \to \eta_c + \gamma) \cong \frac{3}{3} \alpha Q^2 \cdot \mu_c^2 \cdot \frac{p_\gamma{}^3}{m_c{}^2} \qquad (4.3)$$

the charmed quark's magnetic momentum $\mu_c \sim 1/m_c$ is small and if $m_{\eta_c} \cong m_{\psi_c}$ there is not much phase space left (73a,73e). Nearly monochromatic photons are expected through the decays

$$(\psi,\psi') \rightarrow \eta_c + \gamma$$
$$\hookrightarrow \gamma\gamma$$

However, if there are substantial other one-photon decays of the ψ's it may not be possible to get sufficient resolution in ψ-energies in order to detect η_c. In hadronic reactions η_c production cross section is presumably not too different from that of ψ_c. Primakoff photoproduction(73d) (Fig.63) at high energies off a nucleus with charge Z depends on the unknown quantities $\Gamma(\eta_c \rightarrow \gamma\gamma)$ ($\sim$ 50 kev) and the nuclear form factor F(t). A first estimation of the differential cross section is represented in Fig.64 (73d). The total cross section is expected to rise

$$E_\gamma = 100 \text{ Gev} \qquad \sigma(\gamma \rightarrow \eta_c) \sim 380 \text{ nb}$$
$$E_\gamma = 200 \text{ Gev} \qquad \sigma(\gamma \rightarrow \eta_c) \sim 1.6\ \mu\text{b}$$

with growing initial CM-energy (Fig. 65)(73a)

c) Decay of η_c

Estimated decay modes are : $\eta_c \rightarrow$ hadrons, $\gamma\gamma$, $\gamma + \pi\pi$. Charmonium (see section 4.2.3) with two-gluon exchange only predicts

$$\Gamma(\eta_c \rightarrow \text{hadrons}) \cong \text{a few Mev} \qquad (4.4)$$

Hadronic final states have quantum numbers $J^{PC} = 0^{-+}$, $G = +1$, $I = 0$. Conventional hadronic 2γ decays have widths : $7\ ev \lesssim \Gamma\ (h \to 2\gamma) \lesssim 20\ kev$ (Table XIV). Thus

$$\Gamma(\eta_c \to \gamma\gamma) = \frac{32}{9} \left(\frac{m_{\eta_c}}{m_h}\right)^3 \Gamma(h \to \gamma\gamma) \simeq 24 - 340\ kev \quad (4.5)$$

is the order of magnitude. Charmonium estimations give a few keV. The small decay width for $\eta \to \gamma + \pi\pi$ suggests that $\eta_c \to \gamma + \pi\pi$ is not too significant[73e].

4) The existence of ψ' at 3700 Mev forces the lowest charmed hadron (with one charmed quark and one or several conventional quarks) above 1850 Mev in mass. $c\bar{c}$-threshold, expected in the range 3.7 - 4.7 Gev, thus suggests lowest mass charmed-mesons in 1850 - 2350 Mev (Table XI). In analogy to uncharmed hadron spectroscopy one expects linear Regge trajectories in the charmed world ; conjectured forms are[65e,66] :

$$\alpha_{\substack{\text{hidden}\\\text{charm}}}(m^2) \cong -3.8 \pm \tfrac{1}{2}\, m^2 \quad (4.6)$$

$$\alpha_{\text{charm}}(m^2) \cong -2.8 + \sqrt{\tfrac{1}{2}} \cdot m^2 \quad (4.7)$$

5) Defining the photon vector-meson coupling constants via the field current identity

$$j^{\mu}_{em}(x) = -e \sum_V \frac{m_V^2}{2\gamma_V} V^{\mu}(x) \quad (4.8)$$

the predictions of SU(4) with ideal mixing

$$\gamma_\rho^{-2}:\gamma_\omega^{-2}:\gamma_\phi^{-2}:\gamma_\psi^{-2}:\gamma_{\psi'}^{-2} = 9:1:2:8:8^{(73d)} \quad \text{(Theory)} \tag{4.9}$$

seem to be in disagreement with the experimentally found values :

$$\gamma_\rho^{-2}:\gamma_\omega^{-2}:\gamma_\phi^{-2}:\gamma_\psi^{-2}:\gamma_{\psi'}^{-2} \cong 9:1:2:2:1 \quad \text{(Experiment)} \tag{4.10}$$

The exact values are[73d] :

$$\begin{aligned} \frac{\gamma_V^2}{4\pi} &= 0.66 \pm 0.05 \quad \text{for } \rho \\ &= 4.8 \pm 0.05 \quad \omega \\ &= 2.84 \pm 0.05 \quad \phi \\ &= 2.80 \pm 0.5 \quad \psi \\ &= 6.60 \pm 1.2 \quad \psi' \end{aligned} \tag{4.11}$$

Discrepancy between theory and experiment is not necessarily a threat to the charm scheme. Possible ways out are :

i) ρ,ω,ϕ,ψ are expected to belong to the same SU_4 quark model multiplet. The large effective masses of the charmed constituents suggest strong SU_4-breaking which might be the cause of the considerable deviations.[5]

Comparison of the ϕ and ω leptonic decay widths suggests that $(\gamma_V/m_V)^2$ is the relevant coupling constant instead of γ_V^2 [65b].

ii) Phase space corrections in leptonic ψ-decay derived by noting that the coupling of the photon depends on the velocity of the quarks in the bound state: (73d)

$$\Gamma_e^{obs} = \Gamma_e^{SU_4}\left(1-\frac{4m_c^2}{M_\psi^2}\right)^{3/2} \qquad (4.12)$$

iii) The above values γ_V^2 do not take into account the bound state nature of $c\bar{c}$ - the effect of $|\psi(0)|^2$ (109ℓ).

6) What are the sizes and important radiative decay modes of ψ : $\psi_c \to \gamma$ + hadrons ? We distinguish :

i) $\psi_c \to \gamma$ + hadrons (with charmed quarks)

ii) $\psi_c \to \gamma$ + hadrons (no charmed quarks)

(Decays like $\psi'_c \to \psi_c + \gamma$ are forbidden by C-conjugation). Within the first group the decay widths depend considerably on the quark content of the charmed hadrons. Supposing

$$\begin{aligned} \eta_c &\equiv \{c\bar{c}\} & M_{\eta_c} &\sim 3 \text{ Gev} \\ X_c &\equiv \tfrac{1}{2}\{p\bar{p} + n\bar{n} + \lambda\bar{\lambda} + c\bar{c}\} & M_{X_c} &= 0.958 \text{ Gev} \\ \eta'_c &\equiv \frac{1}{\sqrt{12}}\{p\bar{p} + n\bar{n} + \lambda\bar{\lambda} - 3c\bar{c}\} & M_{\eta'_c} &= 2.7 \text{ Gev} \end{aligned} \qquad (4.13)$$

the estimated widths are (65e,66):

$$\begin{aligned} \Gamma(\psi_c \to \gamma + \eta_c) &\sim 20 - 50 \text{ kev} \\ \Gamma(\psi_c \to \gamma + x_c) &\sim 20 \text{ Mev} \\ \Gamma(\psi_c \to \gamma + \eta'_c) &\sim 1.5 \text{ Mev} \end{aligned} \qquad (4.14)$$

Transition to states with ordinary <u>and</u> charmed quarks is far too big and some suppression mechanism has to be at work in such a picture. Mixing schemes can reduce the large values by a factor 10^{-2} (65e). Dominant contribution to radiative decays is expected from transition between different $c\bar{c}$ - bound states (Fig. 62). Using Charmonium as a guide (see section 4.2.3.), one expects $\sim 1/3$ of all ψ_c'-decays by the emission of γ s. The second group of transition into ordinary hadrons as for instance

$$\psi_c \to \gamma + \pi \ , \ \gamma + \pi\pi \ , \text{ etc.}$$

is expected to be suppressed since there are no charmed quarks in the final state . ϕ ($\equiv \lambda\lambda$) - decay strongly suggests such a conclusion and Charmonium calculations agree. Forbidden decay modes as eg. $\psi_c \to \pi^0 + \gamma$ might however show up at a similar level to the others in view of their large phase space available.(65b) .

7) Early conjectures(65) expected the K-meson (strange particle) rate to increase considerably whilst going through the resonance peak or charm threshold since in the Bjorken-Glashow-Iliopoulos-Maiani scheme(67), the Cabibbo-favoured weak transition ($\propto \cos\theta$) of the charmed quark is to a strange quark. A recent more careful analysis(78b) now shows that this argument is not correct and that charm production will be difficult to recognize since no dramatic signal may be expected in $e^+e^- \to$ hadrons. However, just above charm threshold the inclusive μ/h or e/h ratio should show a distinct rise in $x \equiv 2E/\sqrt{q^2}$(74).

The experiment at $E_{CM} = 4.8$ Gev shows ~ 20 events: $e^+e^- \to e^{\pm}\mu^{\mp} + X^0$ where $(E_\mu + E_e) \sim \frac{1}{2} \cdot E_{CM}$ and the $(e\mu)$ invariant mass distribution resembles a recoil distribution . The relative angle between the outgoing $e\mu$ -pair is $\sim \frac{3\pi}{4}$ (13,10m)

8) The determination of the leptonic decay width, saturating Weinberg's first sum rule by pole dominance approximation in SU(4) , predicts : $\Gamma(\psi \to e^+e^-) = 1\text{-}5$ kev. - in good agreement (75). We also mention the purely algebraic approach using current commutators of the SU(4) charges V_i : $[V_i, V_j] = i\, f_{ijk} \cdot V_k$ $j = 0, \ldots, 15$. These lead to sum rules whose saturation with vector meson states predicts decay widths which do not contradict the experimental values (76).

9) Aside from the charmed quark charge assignement $Q = 2/3$, there is an alternative : $Q = -4/3$, $S = I = 0$ (77).
Consequences are :

- a realistic value for the integral over the resonance cross section : $\Gamma_e = 5.7$ Kev . Note $(\gamma_V/m_V)^2$ is taken as the relevant γ-ψ coupling constant.
- a hadron to μ-pair ratio : $R \equiv \sigma_h/\sigma_\mu = 7 \cdot \frac{1}{3}$,
- the existence of doubly (triply) charged charmed mesons (baryons) and preferable transitions $c \to n + \ell^- + \bar{\nu}$ (whereas $c \to \lambda + \ell^+ + \nu$ or $n + \lambda^+ + \nu$ in the conventional scheme). This latter feature could explain the <u>non-dominance of strange hadrons</u> in the present experimental results around the resonances.

10) Supposing charmed quarks do exist - what are the further consequences ? The decay rates of charmed hadrons are estimated, supposing $c \to n + \lambda^+ + \nu$ or $\lambda + \ell^+ + \nu$ and neglecting terms with $\sin^2\theta_c$, to be of the order

$$\Gamma(h_c \to \ell + \nu + \text{hadrons}) \simeq \frac{G_F^{\,2}\cdot m_c^{\,5}}{196\cdot\pi^3} \approx 10^{12}\,\text{sec}^{-1} = 6.6\cdot 10^{-10}\,\text{Mev}. \qquad (4.15)$$

$$\Gamma(h_c \to \text{hadrons}) \approx 10^{13}\,\text{sec}^{-1} = 6.6\cdot 10^{-9}\ \text{Mev} \qquad (4.16)$$

In e^+e^- annihilation a dramatic threshold effect is expected; jet structure should become more pronounced in this region. In deep inelastic ν $(\bar{\nu})$ p scattering the additional contributions (due to charm) in the total cross sections σ^ν and $\sigma^{\bar{\nu}}$ are expected to be the same, whereas the fractional increase is roughly three times larger due to $\sigma^{\bar{\nu}}/\sigma^\nu \approx \frac{1}{3}$ at high energies (65e,66). Note however that $\frac{\sigma(\nu N \to \nu N + \psi_c)}{\sigma(\nu N \to \nu X)} \approx 10^{-7} - 10^{-4}$ (65b) .

In deep inelastic ep (or μp) scattering, scaling violation at small x is expected. In photoproduction, $\lesssim$ 1% of the total cross section should be due to charmed particle production. In hadronic collisions, the inclusive cross section due to charmed particle production is estimated to be of the order of 1-10 μb ; large p_t-measurements might be one of the promising places to look for (65e) Two-body quantum number exchange reactions were found to be unfavorable for charm searches (78a) .

4.1.2. Colour

The extension of the usual Gell-Mann-Zweig quark scheme (G-Z)[79] to $SU_3 \times SU_3^c$ by the introduction of a new quantum number leads to 3 x 3 = 9 quarks. The new quantum number, colour (with three eigenvalues attached to each u,d,s quark), gives anti-symmetric quark wave functions and permits quarks with integer charge.

The G-Z triplet scheme with non-integer quarks[80a] and a colour singlet photon is distinct from the Han-Nambu scheme with integer values of the quark charges and the appearence of coloured hadrons and a photon[81].

$$(3,\bar{3}) \times (\bar{3}, 3) = \underbrace{(1,1) + (8,1)}_{\text{colour singlet}} + \underbrace{(1,8) + 8,8)}_{\text{colour octet}} \qquad (4.17)$$

The G-Z scheme does not help to solve the ψ-mysteries but may be useful in the context of the $c\bar{c}$-interpretation. By assumption, the lowest baryon and meson states are colour singlets with respect to SU_3^c but large mass hadron states are supposed to carry colour with charge :

$$Q = (I_3 + \tfrac{1}{2} Y) + (I_3^c + \tfrac{1}{2} Y^c) \qquad (4.18)$$

The photon has components in both (1,8) and (8,1) .

Since all ordinary hadrons are SU_3^c-singlets, $\psi \equiv \psi$(colour) is absolutely forbidden to decay into ordinary hadrons (only), via strong interactions. This is the colour explanation for the narrow width of ψ (Fig.66) . The $\psi \equiv \psi$(colour) coupling to leptons is conventional (via single photon exchange) and it occurs because of mixing with the photon. There exists a wealth of different schemes along this line of reasoning which we do not intend to present here but rather refer to the review by Greenberg[82]. We limit ourselves to list the proposed assignments (Table XV) and indicate some of the characteristics and difficulties[83]:

1) Decays like $\psi \to \gamma$ + hadrons are not forbidden (since the photon carries colour away) and the multiparticle final states could even have a large neutral energy fraction. However, rough estimates give a decay rate which is far too large - a problem familiar to the charm decay scheme where Zweig's selection rule still might help. Some suggested ways out are[82] :

(i) Radiative M_1 decays of vector mesons demand that the coloured mesons' magnetic moment be small ; $\mu_c/\mu_p \lesssim 1/20$ which in turn demands that the coloured baryon masses $\gtrsim$ 26 Gev . However, this argument would not work for E_1 transitions.

(ii) Dynamical effects suppress emission of photons with $q^2 = 0$ whereas coupling of virtual coloured photons to highmass coloured hadron states (as ψ) is not suppressed ; thus, the γ-ψ coupling constant $f_{\gamma\psi}(q^2)$ would be strongly q^2-dependent.

(iii) In the quark model, calculation of the transition matrix elements involve the following assumptions : the space wave functions are universal for a super-multiplet and the matrix elements are not momentum dependent. These might fail for ψ .

(iv) All transitions between coloured and non-coloured states occur via an SU_3-singlet with semi-strong interaction of strength : $(g^2/4\pi)_{\text{semi-strong}} \approx (0.1\text{-}0.01)\cdot(g^2/4\pi)_{\text{strong}}$. In any case, a photon should be found in a large fraction of the "hadronic" decays.

2) The existence of two narrow resonances ψ and ψ' and their decay impose restrictions. Their photon couplings demand them to be in a colour octet. Some of the more specific models can provide explanations for the existence of two narrow resonances, but suffer from the disadvantages of accepting an unreasonably large colour breaking and/or demanding the existence of their charged partners. The ψ-enhancements then might even resolve themselves in two resonances, due to vector meson mass differences of electromagnetic origin.

3) No additional narrow vector mesons are expected in such a scheme; wider ones may well exist[84].

4) Some investigations[85] assume ψ and ψ' to be in $(1,8)+(8,8)$: an octet in colour and a singlet-octet mixture in SU_3 . In analogy with ω and ϕ one may identify : $\psi \equiv \omega_c$, $\psi' \equiv \phi_c$. This predicts the vector meson-photon coupling constants as

$$f_\rho^{-2} : f_\omega^{-2} : f_\phi^{-2} : f_\psi^{-2} : f_{\psi'}^{-2} = 9 : 1 : 2 : 8 : 4 . \tag{4.19}$$

The other members of the octet cannot couple to the photon and therefore are not seen in e^+e^- - initiated experiments. In Fig.67 , we show an anticipated level structure for colour under the above assignment of ψ s. There is full nonet symmetry for the vector states and ε_c, δ_c, η_c are taken to be pure $u\bar{u} + \bar{d}d$. For comparison a charm spectrum is drawn[85d].

5) Doubly charged states with narrow widths in the 3-4 Gev region are expected (but have not been found so far)[85].

6) The decay $\psi' \to \psi + \pi\pi$ is a priori not suppressed in the colour approach; however, some of the specific models do offer suppression mechanisms such as Zweig selection rule , isospin conservation and SU_3-singlet transition (via semi-strong interactions) etc.[82,83]

7) Deep inelastic ep scattering should show (temporary) scaling violations due to the production of high mass coloured hadrons. One expects an increase of about 100 % in νW_2[65e].The asymptotic hadron to μ-pair production ratio changes from $R_\infty = 2$ (G-Z) to $R_\infty = 4$ (H-N)[4a]. Such a scheme also predicts an improved value for the π^o-life time via Adler anomaly.

8) Using the generalized Weinberg sum rules, the leptonic decay widths are estimated to be : $\Gamma(\psi \to e^+e^-) \sim 9$ kev[86].

9) A modification of the colour scheme is obtained by <u>adding another triplet of heavy quarks to the Han-Nambu scheme</u>[87]; thus the charge assignment is

$$Q = \begin{pmatrix} p_1 & p_2 & p_3 & p_4 \\ n_1 & n_2 & n_3 & n_4 \\ \lambda_1 & \lambda_2 & \lambda_3 & \lambda_4 \end{pmatrix} = \begin{pmatrix} 0 & 1 & 1 & 0 \\ -1 & 0 & 0 & -1 \\ -1 & 0 & 0 & -1 \end{pmatrix} \qquad (4.20)$$

The fourth triplet is restricted to transform under SU_3 like a triplet and under SU_3^c like a singlet. A large mass difference is postulated between the Han-Nambu and the additional quarks. This model is a particular form of a coloured charm model. Particles are assigned to the three neutral states that couple to the photon as follows :

$$\begin{aligned} \psi(3.1) &= \lambda_4\bar{\lambda}_4 \cdot \cos\delta + \sqrt{\tfrac{1}{2}}\,(p_4\bar{p}_4 + n_4\bar{n}_4) \cdot \sin\delta \\ \psi(2.7) &= \sqrt{\tfrac{1}{2}}\,(p_4\bar{p}_4 + n_4\bar{n}_4)\cos\delta - \lambda_4\bar{\lambda}_4 \cdot \sin\delta \qquad (4.21) \\ \psi(4.1) &= \sqrt{\tfrac{1}{2}}\,(p_4\bar{p}_4 - n_4\bar{n}_4). \end{aligned}$$

The particular features here are : <u>$\psi(4.1)$ has two charged isospin partners degenerate in mass. $\psi(3.1)$ and $\psi(3.7)$ have $G = -1$, whereas $\psi(4.1)$ has $G = +1$</u>[87]. <u>$R_\infty = 6$</u> .

<u>4.1.3. Other Ideas</u>

Among the suggestions for new quantum numbers we pick out the ones which seem interesting to us.

Supposing that there is a new additive quantum number in strong interactions[88], one might try schemes similar to K^o, $\bar{K}^o$[127]. These would result if there are two J^o s

$$J_o = J_1 + iJ_2 \qquad \bar{J}_o = J_1 - iJ_2 \qquad (4.22)$$

defined as linear combinations of two hermitian fields J_1 and J_2. The observed ψs then are linear combinations of J_1 and J_2. In this case, one has a new additive quantum number t (= ± 1,0) which is conserved in strong interactions : $t = +1$ for J^o and $t = -1$ for $\bar{J}^o$, $t = 0$ for the pre-ψ-particles. Such assumption makes the interaction ψ-h-ψ possible. It is expected to be strong. We mention here a similar attempt in weak interactions[129] and refer to section 4.2.7. for further consequences and details of such quantum number assignment[127].

The ad hoc introduction of new quantum numbers obviously can explain the small hadronic ψ-widths by selection rules which forbid hadronic decays as a whole. One might try quantum number assignments such as[89] :

$$\begin{array}{lll} \xi = 0 \text{ hadrons} \quad \text{or} & \xi = 0 \text{ hadrons} \quad \text{or} & \xi = 0 \text{ hadrons} \\ \xi = 1\ \psi, \psi' \ldots & \xi = 1\ \psi & \xi = 1 \text{ leptons} \\ & \xi = 2\ \psi' & \xi = 2\ \psi, \psi' \ldots \end{array} \qquad (4.23)$$

These can explain suppression of certain production and decay channels which, however, are only specified for a few particular cases : radiative decay modes, $\psi' \to \psi + \pi\pi$, etc. Further consequences are : $\Gamma_{\psi'} \sim (5-10) \cdot \Gamma_{\psi}$, the absence of a similar type of resonance below 2.5 Gev(89a).

ψ-particles are put in a new class of objects; one might question such a point of view by saying, for instance, that the decay $\psi' \to \psi + \pi\pi$ seems to be strong (see section 4.2.1).

A proposal(90), somewhat similar to the charm scheme (but clearly distinct), also introduces a new quark $\chi(Z)$ which carries the new quantum number $Z \equiv$ paracharge. Its group theoretical structure is $SU_3 \times SU_1(Z)$ and the quantum number assignment to the quarks (p, n, λ, χ) is as in Table XVI . $\psi(3.1)$ is accomodated in the $\underline{15} + \underline{1}$ representation of SU_4 containing the usual SU_3 octet of vector mesons of the ρ-family. $\psi'(3.7)$ belongs to another SU_4 15-plet which contains the $\rho'(1600)$ etc. In addition to the old nonet seven new vector mesons are added : (D^+, D^o), $(\bar{D}^o, D^o)$, S, $\bar{S}$ and P_1 . The essential characteristics of this scheme were found to be :

(i) There should be <u>two further resonances in the ranges 4.1 - 4.3 Gev and 5.0 - 5.2 Gev</u> .

(ii) The dominant decays of $\psi(3.1)$ are into $\gamma + h(Z = \pm 1,0)$. Therefore monochromatic higher energy photons both at the resonance peaks and above threshold of new particles are expected. Many of them still have to be detected and all of them are predicted to be electromagnetically unstable. Decays like $\psi(3.1) \to n\pi + \gamma$ are rare. Note, purely strong decays are negligible - which is in disagreement with the experiment! Estimated values of the main decays are

$$\begin{aligned} \Gamma(\psi \to S_+^o + \gamma) &\simeq 20 \text{ kev} \\ \Gamma(\psi \to \eta\gamma + \eta'\gamma) &\simeq 45 \text{ kev} \end{aligned} \tag{4.24}$$

(iii) ψ'-decays are :

$$\begin{aligned} \psi(3.7) &\to \psi(3.1) + 2\pi \quad \text{(dominating)} \\ &\to S_-^o\ (PS) + 2\pi \quad \text{(still large)} \\ &\to \eta\gamma + \eta'\gamma \\ &\to S_+^o(3.1) + \gamma \end{aligned} \tag{4.25}$$

The usual electromagnetic current must be modified by the addition of an anomalous piece that enables transitions with $\Delta Z = \pm 1$, $\Delta I = \Delta Y = 0$. Its contribution in e^+e^- annihilation is expected to dominate over the normal one at asymptotic energies and thus might be responsible for the asymptotic constancy of the total cross section.

(iv) The large fraction of energy carried away by neutral particles (energetic photons for the initial decay and π^os) provides an explanation for the energy crisis. In addition $K^-/\pi^- \simeq 10\%$ at $\sqrt{q^2} \simeq 4.8$ Gev[90].

We mention suggestions which go beyond one additional (heavy) quark.

One might assume the existence of two heavy charmed quarks c and f whose masses are split by weak interactions only[91]. A symmetry of strong and electromagnetic interactions (analogous to isospin), called X-spin, then transforms one into the other. Vector mesons formed of c and f quarks will fall into families with X-spin 0 and 1. Since the photon carries X-spin zero, the X = 1 family cannot be detected in e^+e^- annihilation. Combinations of c and f quarks with ordinary quarks, for instance $\bar{c}p$ or $\bar{f}p$ states, will form families with very close degenerate masses, split by weak interactions only. A test of such scheme is : weakly decaying particles with masses close to 3.1 Gev produced in pairs in hadronic collisions.

One might go beyond two and try schemes with three additional (heavy, charmed) quarks[92]; thus

$$(p, n, \lambda \; ; \; p', n', \lambda')$$

where the last three quarks have the same quantum numbers respectively as the first three, except they have $C = +1$ instead of zero. The narrow resonances found in the e^+e^- channel and in $pp \to \tilde{\psi} + X$ are not the same in this model. They are identified with

$$\rho_2 = \tfrac{1}{2}\,(p'\bar{p}' - n'\bar{n}') = \psi\ (3.1) \quad (\text{in } e^+e^-)$$

$$\omega_2 = \tfrac{1}{2}\,(p'\bar{p}' + n'\bar{n}') = \tilde{\psi}\ (3.1) \quad (\text{in } pp) \qquad (4.26)$$

$$\phi_2 = \lambda'\bar{\lambda}' \quad (\text{beyond } 4.1 \text{ Gev})$$

$\psi'(3.7)$ and $\psi''(4.1)$ are considered to be radial excitations of $\psi(3.1)$.

As an alternative to this scheme two fundamental building blocks were proposed : the usual SU_3-triplet (p,n,λ) and an SU_3-antitriplet (p',n',λ') of heavy quarks (Fig. 68) which carries a new quantum number $H \equiv$ Heaviness [93]. In addition all quarks are supposed to be "coloured". The identification with narrow resonances is

$$\psi\ (3.1) = (1/\sqrt{3})\ \{p'\bar{p}' + n'\bar{n}' + \lambda'\bar{\lambda}'\}$$

$$\psi'(3.7) = (1/\sqrt{6})\ \{p'\bar{p}' + n'\bar{n}' - 2\lambda'\bar{\lambda}'\} \qquad (4.27)$$

$$\psi''(4.1) = (1/\sqrt{2})\ \{p'\bar{p}' - n'\bar{n}'\}$$

Some of the consequences are : $R_\infty = 5$. The existence of charged partners $\psi^{\pm}(4.1)$, and of four strange ψ s around 3.8 Gev is postulated. The decay $\psi' \to \psi + \eta$ is allowed (but SU_3-forbidden in the charm scheme) as well as $\psi' \to \psi + \pi\pi$. There are no $K\bar{K}$ or $K^*\bar{K}^*$ decays of ψ; this however is a more general consequence of exact SU_3-symmetry [65c].

All radially excited states of the heavy quarks are expected beyond 4.1 Gev and very wide.

Another suggestion[94] introduces, aside from the usual (p,n,λ) quarks, three new ones which are arranged as

$$(p,n,\lambda;\ g^{+},g^{o},g^{-})$$

In this six quark model weak neutral currents can be constructed in the usual way. The strangeness changing part is proportional to the difference of the quark masses $(\frac{\Delta m}{M_W})^2$ and can be suppressed. Mesonic, electrically neutral, bound states $(\bar{g}^{+}, g^{+})$, $(\bar{g}^{-}, g^{-})$ etc. could be produced through electromagnetic or weak interactions and they obviously could undergo strong interaction decays with broad widths.

The observed widths have the size of purely electromagnetic decays which leads to the postulate of a new quantum number g = gentless that is conserved for strong interactions. In analogy to isospin it is supposed to have three eigenvalues

$$\begin{aligned} g &= +1,\ g_3 = (+1,\ 0,\ -1) \quad \text{for} \quad (g^{+},g^{o},g^{-}) \text{ - quarks} \\ g &= 0,\ g_3 = 0 \quad\quad\quad\quad\quad\ \ \text{for} \quad (p,n,\lambda) \text{ - quarks} \end{aligned} \tag{4.28}$$

Electromagnetic interactions do violate g-spin just as they violate isospin conservation. Predicted consequences are : <u>in photoproduction, charged counter parts to the neutral narrow resonances should be seen, for instance $(\bar{g}^{o}, p^{+})$ etc.</u>[94] .

A more recent paper[95] points to the group G(2) which contains SU_3 as a sub-group. The 7-dimensional representation of G(2) breaks up as :

$$7 = 3 + \bar{3} + 1 \tag{4.29}$$

and is interpreted as a triplet of triality $t = \pm 1, 0$ - triplet, antitriplet, singlet - all with baryon number B = 1/3. Mesons are identified with the 14-plet of

$$7 \times 7 = 1 + 7 + \boxed{14} + 27 \tag{4.30}$$

and baryons with the 14-plet of

$$7 \times 7 \times 7 = 1 + 4 \cdot [7] + 2 \cdot \boxed{[14]} + 3 \cdot [27] + 2 \cdot [64] + [77]. \tag{4.31}$$

the charge is defined as

$$Q = I_3 + Y/2 + t/3 \tag{4.32}$$

Predictions are : another vector meson $\psi^+(3.1)$ aside from $\psi^o(3.1)$ and $\psi^o(3.7)$ and a triplet of pseudoscalar mesons. A triplet and an anti-triplet of baryon should exist. The asymptotic hadron to μ-pair ratio is : $R_\infty = 2$[95].

A dynamical SU(3) scheme based on the group embedded in SU(3) → SU(3,1) is a further recently proposed possibility. It leaves several options for ψ assignment[96].

Some of the main characteristics of these schemes with one or several additional quarks are the prediction of many new particles - carriers of new quantum numbers - as well as new narrow resonances, some of them with charge.

4.1.4 Lepton Quarks

We report in this section on attempts which introduce the possibility of leptonic quarks, or which are based on higher symmetry schemes unifying weak and strong interacting particles in one framework.

We start with a system in weak interactions which supposes a weak U_3 in analogy to the strong SU_3 in hadron physics [97]. The additive quantum numbers of this "weak group" are Q, I_3, L in analogy to the strong SU_3 ; the quantum number L takes the rôle of hypercharge. There are three fundamental "weak quarks" but four candidates : e^-, ν_e, μ^-, ν_μ . This scheme abandons the distinction of two neutrinos and classifies the fundamental U_3 triplet as represented in Fig.69 . The leptonic quantum number assignment (Table XVII) is according to Konopinski-Mahmoud[97b] . Only one ν is neede since the helicity assignment of the neutrinos acts as another quantum number. The charge is :

$$Q = I_3 + \frac{1}{4}\,(L-12\cdot C) \qquad\qquad (C = \sum_i^3 a_i) \qquad\qquad (4.33)$$

The a_i's are the eignevalues of the diagonal part of the generators A_{ij} of the group. One may extend this group to SU_4, introducing a new lepton quark ℓ with the quark number assignment given in Table XVIII.

Thus the new "weak quark" has electric charge Q = 2, isospin I = 0 and L = -1. Using the reduction $\underline{4}\times\underline{4}^* = \underline{1} + \underline{15}$, there are three composite particles with $I_3 = L = Q = 0$ in the center of the 15- representation and a singlet in the 1 - representation, thus four uncharged particles. It was suggested that they should be identified with : $\gamma, \psi, \psi', \psi''$. Such picture predicts the following particularities :

1) A rise in R when weak quarks ℓ^{++}, ℓ^{--} are produced, a rapid increase in 4-prong and 6-prong events is expected due to their large charge. Jet structure should appear.

2) Further resonance enhancements are expected in $e^+\mu^-$ and $e^-\mu^+$ final state channels due to generation of intermediate states with ($L = \pm 2, I_3 = \mp \frac{1}{2}, Q = 0$) in the 15 representation.

3) $\ell^{++} \to e^{\pm}\pi^{\pm}$ are possible and a bump in the $e^{\pm}\pi^{\pm}$ mass plots should turn up at $m_\ell \approx 2$ Gev.[97]

The idea of classifying leptons in the same way as for hadrons, by the introduction of leptonic quarks ≡ lorks, has been proposed and many new (heavy) leptonic states were predicted on the basis of such assumption. Single heavy lepton production is prohibited in lowest order and only (higher threshold) pair production of charged heavy leptons is allowed [98].

We mention in this context a similar attempt [99] which proposes the quantum number assignment shown in Table XIX whereby $Q = I_3 + (\frac{\ell+S}{2})$.

In many of its characteristics, the following model stands quite close to color and charm schemes. However its attempt to unify leptons and hadrons leads us to present it in this section.

So far a clear distinction between hadrons and leptons was taken for granted. To overcome this (perhaps artificial) separation, the unification of hadronic quarks with leptons was proposed [100]. Since quarks may carry color-degree of freedom and a heavy charmed quark was postulated, one might go further and group these fermionic constituents in a 16-fold of 4-component fields transforming as $(4,\underline{4})$ of $SU_4 \times SU_4^c$. The fourth color thus is the lepton number L

$$F = \begin{pmatrix} p_1 & p_2 & p_3 & p_4(=\nu_e) \\ n_1 & n_2 & n_3 & n_4(=e^-) \\ \lambda_1 & \lambda_2 & \lambda_3 & \lambda_4(=\mu^-) \\ c_1 & c_2 & c_3 & c_4(=\nu_\mu) \end{pmatrix} \begin{matrix} \Big\} \text{ I - spin} \\ \\ \text{Strangeness} \\ \text{Charm} \end{matrix} \qquad (4.34)$$

← Four colors →

This multiplet is used to construct a theory analogous to the gauge theory unifying weak and electromagnetic interactions which now also includes strong interactions. The assumed underlying group structure is

$$G = SU(4)_L \times SU(4)_R \times SU(4')$$

where the last one corresponds to the 15 universally coupled color gauge mesons. They are indicated by the matrix

$$\left(\begin{array}{c|c} & X^{o} \\ V(8),\ S^{o} & X^{-} \\ & X^{-'} \\ \hline \bar{X}^{o}\ \bar{X}^{-}\ \bar{X}^{-'} & S^{o} \end{array}\right) \qquad (4.35)$$

V(8) represents an SU_3^c color octet of vector gluons ($m_V \approx$ 10 Gev) which mediate "conventional" strong interactions of quarks. S^o is a singlet which gives rise to leptonic as well as semi-leptonic interactions ($m_{S^o} \gtrsim 10^3$ Gev), while not affecting purely leptonic or purely hadronic reactions. The X-particles induce semi-leptonic reactions in lowest order as follows : $X \rightarrow \ell + \bar{q}$ or $\ell + \bar{\ell} \rightarrow q + q^-$ ($m_x^2 \gtrsim 10^4$ Gev). The coupling strength in this model is universal $f^2/4\pi \approx 1$ and the contributing strength of the various gluons is modified by the choice of their masses.

General consequences of this scheme are :

(i) There exists a new class of interactions between lepton-baryon and lepton-lepton which is different from the familiar weak and electromagnetic interactions.

(ii) Baryonic quarks can transform into leptons. Thus lepton and baryon quantum numbers are no longer conserved, but fermion number $F = B + L$ still is. As a result unstable integer-charged quarks exist and the proton, although long-lived, becomes unstable.

Among the fifteen gauge particles presented above, the fundamental octet V(8) (which provides the "glue" for the quark binding) should be produced at presently available energies. It is predicted to occur in the mass region 3-5 Gev and splits into

$$\begin{array}{ll} \text{4 charged members} & : (V_\rho^\pm, V_{K^*}^\pm) \\ \text{4 neutral members} & : (V_3, V_8, V_{K^*}^o, \bar{V}_{K^*}^o) \end{array} \tag{4.36}$$

These particles would be produced in pairs (or in association with other color octet baryons or mesons) in collisions with energetic hadrons ; the two neutral members V_3 and V_8 or their linear combination

$$U^o = \frac{1}{2} (\sqrt{3}\, V_3 + V_8) \ , \ V^o = \frac{1}{2} (\sqrt{3}\, V_8 - V_3) \tag{4.37}$$

can be produced singly in e^+e^- annihilation and all photon induced reactions.

We should mention that spontaneous symmetry breaking, while giving masses to the gauge bosons, induces mixing between several of them so that physical gauge particles are, in general, mixtures of the canonical gauge particles. The most interesting candidates to be interpreted as ψs are

$$U = \cos\xi . \tilde{U} + \sin\xi . V^{o} \qquad (4.38)$$

$$V = -\sin\xi . \tilde{U} + \cos\xi . V^{o} \qquad (4.39)$$

$$\tilde{U} \simeq \frac{\sqrt{3}\, f \cdot U^{o} - g \cdot W}{\sqrt{3f^2 + 2g^2}} \qquad W \equiv W_L^3 + W_R^3 - \sqrt{\frac{2}{3}}\left(\frac{g}{f}\right) \cdot S^{o}$$

Some of their decay characteristics are :

(i) Leptonic decays $U,V \to e^+e^-$ can permit unambiguous determination of the fundamental effective coupling constant f. If $\frac{f^2}{4\pi} \approx 7\text{-}14$ and $\cos^2\xi \sim 0.5\text{-}1$ then

$$\Gamma(U,V \to e^+e^-) = \left(\frac{2}{3} \cdot \frac{e^2}{f}\right)^2 \cdot \frac{M_{U,V}}{12} \cdot \cos^2\xi \simeq 5 \text{ Kev} \qquad (4.40)$$

(ii) Typical hadronic decays are :

$$U \to 3\pi, \pi\pi\omega, \rho\pi, K^*\bar{K}, B\pi \quad \{ O(H_8') , O(\alpha) \}$$

$$U \to \pi^+\pi^-, 4\pi, K\bar{K}, \eta\pi\pi, \rho\rho \quad \{ O(\alpha) \} \qquad (4.41)$$

with

$$\Gamma(U \to \text{hadrons}) \approx 50 - 200 \text{ kev.}$$

(iii) Some of the radiative decay modes are forbidden

$$U \not\to \pi^{o} + \gamma \quad \text{but} \quad U \to \eta'(\text{or } E^{o}) + \gamma$$

$$U \not\to (2n+1) . \pi + \gamma \qquad U \to (2n)\pi + \gamma \qquad (4.42)$$

$$U \not\to \eta + \gamma$$

with

$$\Gamma(U \to h + \gamma) \approx 50 - 100 \text{ keV}. \qquad (4.43)$$

Withing the above framework the following three possible assignments for the ψ s were proposed :

Model I : ψ (3.1) = color-gluon U

ψ'(3.7) = color $(q\bar{q})$-compound $\varepsilon\ C_U$ (4.44)

ψ''(4.1) = $c\bar{c}$-compound : $\emptyset_c$

Some characteristics of this model are :

(h+γ)-decays constitute a significant fraction (20-50%) of all decay modes. The specific modes $(\eta+\gamma)$, $(E^0+\gamma)$ should be appreciable ; monoenergetic γ-rays in the U-mass region are expected. The other charged members of the octet V(8) should be found which could be produced singly in neutrino interactions. Two more particles (one out of U and V and one out of C_U and C_V) are predicted to exist which are related to ψ and ψ'. They ought to exist within 50 to 100 Mev of ψ and ψ'. $\psi' \to \psi + \pi\pi$ is strong, but suppressed by Zweig's rule.

Model II : ψ (3.1) = color gluon U

$$\psi'(3.7) = \sqrt{\tfrac{2}{3}}\ (\omega\ ,\gamma_c) + \sqrt{\tfrac{1}{3}}\ (\emptyset,\gamma_c) \qquad (4.45)$$

$$\psi''(4.1) = -\sqrt{\tfrac{1}{3}}\ (\omega_1,\ \gamma_c) + \sqrt{\tfrac{2}{3}}\ (\omega_8,\gamma_c)$$

The spectrum of this model is similar to the above, except

that doubly charged partners of (ω, χ_c) and $(\emptyset, \gamma_c)$ are expected near 3.7 and 4.1 GeV. Decays $\psi'' \to \psi' + \eta$ $\psi' \to \psi + \eta$ account for a large part of the ψ'' and ψ' decay widths ; thus mono-energetic η's in these two decays are significant characteristics of this model. Zweig's rule suppresses $\psi' \to \psi + \pi\pi$.

Model III :

$$\begin{aligned} \psi \ (3.1) &= \emptyset \ (c\bar{c} \text{ - compound}) \\ \psi' \ (3.7) &= \emptyset' \ (\text{radial excitation}) \\ \psi''(4.1) &= U \end{aligned} \qquad (4.46)$$

This model shows the characteristics of the charm model. The large ψ'' decay-width is a consequence of transitions to lower colored pseudo-scalars : $\psi'' \to (1,8) + \eta\pi$

Note in these models V could have been chosen instead of U.[100]

4.2. Hadrons : Dynamical Schemes

Let us suppose these narrow resonances are of hadronic nature and there are dynamical mechanisms which bring their widths to these small values.

4.2.1 Consequences

Such picture has a few immediate consequences which we would like to present first. If the photon couples to ψ and ψ' 'a la VMD' then vacuum polarization effects [101] can no longer be neglected and the full photon propagator (with vacuum polarization) has to be taken into account according to

$$\frac{1}{q^2} \rightarrow \frac{1}{q^2} \cdot \frac{1}{1-\pi(q^2)} \tag{4.47}$$

An investigation shows that Im $\pi(m_\psi^2) \sim 0.5$. Supposing an unsubtracted dispersion relation for $\pi(q^2)$, the influence of this effect is an asymmetry in the (background) QED single photon exchange amplitude which results in a considerable supression at the lower end of the resonance and a rise at its upper end. The same interference pattern is observed if single photon exchange and a narrow resonance Breit-Wigner form are added [101].

The experiment does show the anticipated pattern.

One of the major questions is : are the new states strongly interacting hadrons ?

The size of σ_{tot} (ψN) and $\sigma(\gamma\to\psi)$ are therefore of importance and should be in the range of strong interaction values : millibarn resp. nano-barn. From photonic meson production one would expect $\sigma(\gamma\to\psi)$ in the micro-barn range; however Table IV shows a constant decrease of the vector meson production rate with increasing mass m_{V_0} .

One way to test this is to relate <u>ψ-photoproduction $\gamma N \to \psi N$</u>[102], via vector meson dominance, to diffractive ψN-scattering and this in turn to the total cross section :[65b]

$$\sigma(\gamma N\to\psi N)=\{\frac{\alpha}{4}\cdot\frac{4\pi}{\gamma_V^2}\}\cdot\sigma_{Diff.}(VN\to VN)=\{\frac{\alpha}{4}\cdot\frac{4\pi}{\gamma_V^2}\}\cdot\{\frac{1}{16\pi b}\}\cdot\sigma^2_{tot}(VN) \tag{4.48}$$

If the photoproduction cross section is in the range $0.2<\sigma(\gamma N\to\psi N)<4\mu b$ a simple estimate predicts the hadronic total cross section in the range : $5mb < \sigma(\psi N) < 10$ mb. The experimental values are around 2 nb for $E_\gamma = 18$ Gev[65b].

This reasoning involves the following assumptions :

(i) γ_ψ is supposed to show little q^2-dependence - is $\gamma_\psi(q^2)$ = const. ? VMD implies the approximation of the scattering amplitude for $\gamma N\to VN$,$T(s;q^2)$, by a pole in the photon mass; this might be wrong, other contributions (and singularities) could equally well be influencial .

(ii) $\gamma(V)N \to VN$ is likely to be diffractive, but it does not <u>have</u> to be.

Model estimates for $\sigma(\gamma N \to \psi N)$ are based on :

a) vector meson dominance supplemented by [102b,102f]

$$\lim_{q^2\to\infty} \frac{\sigma_{tot}(V_1N)}{\sigma_{tot}(V_2N)} = \left(\frac{m_2^2}{m_1^2} \right) \qquad (4.49)$$

Consideration of the ratio $\sigma(\gamma N\to \psi N)/\sigma(\gamma N \to \phi N)$ then predicts : $\sigma(\gamma N\to\psi N) \sim 6$ nb

Note, the relation(4.49) is used in the extended vector meson dominance picture which predicts similar orders of magnitude. It implies scaling behaviour for the diffractive vector meson cross section and has been shown to hold in the dual resonance model [102i].

b) The Bohrradius of Charmonium models [102 f]

$$R_{Bohr} = \frac{n^2}{(\frac{4}{3}\alpha_s)(\frac{m_c}{2})} \sim (0.1-0.5)\cdot n^2 \text{ fm} \qquad (4.50)$$

using $\sigma_{tot} = \text{const}\cdot r^2$ which gives $\sigma(\gamma N\to\psi N) \simeq 0.2-26$nb.

c) a stat. model for the exclusive reaction $pp \to pp+\omega$ [102a] continued in m_ω to the ψ-mass. Such procedure leads to a strong suppression factor estimated as : $h^2 \sim 3\cdot 10^{-4}$. Taking this value to photoproduction :

$$\frac{d\sigma}{dt}(\gamma p \to \psi p) \cong h^2 \cdot \frac{d\sigma}{dt}(\gamma p \to \omega p)\Big|_{m_\omega \to m_\psi} \qquad (4.51)$$

one finds

$$\sigma(\gamma p \to \psi\ p) \sim 0.3\text{-}3\ \text{nb} \qquad (4.52)$$ [102a].

A second characteristic which certainly will give indications of the nature of ψ (hadronic or not ?) are the decays [103]

$$\psi'(3.7) \to \psi(3.1) + X \qquad (4.53)$$

$$\psi'(3.7) \to \psi(3.1) + \pi^+\pi^- \qquad (4.54)$$

Representative experimental curves are drawn in Fig. 19 . Several authors have investigated such decay assuming it to proceed either directly by the coupling $f \cdot \psi'_\mu \cdot \psi^\mu\ \phi_{\pi^+} \cdot \phi_{\pi^-}$ or via an 0^+ intermediate state (ε-meson) through the chain $\psi' \to \psi + \varepsilon$. The earlier model gave a coupling constant $f \simeq 9.2$ [103e] but such model clearly disagrees with the π-mass distribution. The latter model, with ε intermediate state and a scalar local π-ε-π coupling, leads to a wide enhancement(103c). It predicts for the ψ'-ε-ψ coupling: $g_{\psi'\psi\varepsilon}\cdot\psi'_\mu\cdot\psi^\mu\cdot\varepsilon$ the coupling constant value $g_{\psi'\psi\varepsilon} \simeq 1.7$. The analogous values for $\rho'\to\rho+\pi^+\pi^-$ are : $f_{\rho'\rho\pi\pi} \simeq 87$,(103e) $g_{\rho'\rho\varepsilon} \simeq 12$. <u>We therefore conclude that there is a suppression of a factor $\sim$10.</u>If derivative coupling for $\varepsilon\to\pi\pi$ is assumed(103c)

$$\frac{g_{\varepsilon\pi\pi}}{m_\varepsilon}\,\partial_\mu\pi^+\,\partial^\mu\pi^-\cdot\varepsilon \qquad (4.55)$$

as suggested by chiral dynamics and low energy $\pi\pi$- scattering, excellent agreement with the $\pi\pi$ -mass distribution is achieved.Suppression is of the same order of magnitude.

<u>4.2.2. Hadron bound states</u> (104)

Within the framework of the known hadrons the consequences of considering ψ as a $B\bar{B}$ compound have been investigated. The $(\Omega\bar{\Omega})$- system offers itself as a possible candidate since its total mass is bigger than 3100 Mev. The stability of such system is imagined by the interplay of attractive forces due to η and ω and repulsive forces which prevent instantaneous annihilation. Those can be due to short range repulsion or due to angular momentum motion leading to a centrifugal barrier. Within the quantum mechanical framework one suggestion conjectures that ψ is a $^7D_1$ of the $\Omega\bar{\Omega}$-system and such picture can indeed predict the right order of magnitude for the decay width [104 a,b]

$$\Gamma \approx \frac{4}{Z_{\ell n}} p \cdot \frac{v_{\ell}(x)}{\{ 1 + \frac{m^2}{2} x^2 \}^{\frac{1}{2}}} \qquad (4.56)$$

ψ' is identified with $^5D_1$ but is then no longer narrow since it lies above threshold. The 4100 Mev enhancement was predicted before its discovery to be a $^3D_1$ state and to be broad. Quark diagram investigations then predict a characteristic hierarchy of final states composed predominantly of strange particles ($\Xi, \Lambda, \kappa, \Sigma, ..$), mesons and nucleons [104k].

A slightly different suggestion [104h] identifies ψ with a $^3S_1$ bound state and ψ' with $^3D_1$, $^7D_1$ or $^7G_1$ resonance of the $\Omega\bar{\Omega}$ - system and expects the dominant decay modes of ψ to be $\phi\,\phi\,\phi$, $\phi\,\phi\,\omega$, $\phi\eta\eta$ and $\phi\,\eta\eta'$. High multiplicity is a further characteristic.

The small width Γ_{ψ} is explained by three features :

(i) the matrix element of $\psi \rightarrow 3\phi$ decay (ϕ-masses are neglected) is assumed to be

$$|\langle f \mid M \mid i \rangle|^2 \propto \text{const.} \cdot (g^2)^3 \; m_{\Omega}^{-2} \qquad (4.57)$$

(as in positronium annihilation to three photons).

(ii) $m_\emptyset$ however is not negligible and its mass is felt through the 3-dimensional phase space factor

$$PS \propto (m_\psi - 3m_\emptyset)^2 \tag{4.58}$$

(iii) the density of $\Omega\bar{\Omega}$ in ψ is taken into account by the pair's wave function evaluated at its spatial origin : $|\psi(o)|$.

The decay width for $\psi \to 3\emptyset$ is

$$\Gamma_\psi \sim |\psi(0)|^2 \cdot |\langle f|M|i\rangle|^2 \cdot PS = 4.9 \cdot (g^6) \text{ Kev} \tag{4.59}$$

If compared with the experimental value of $\Gamma_{exp} \simeq 80$ kev, it then leads to the coupling constant $g^2 \sim 2.5$ - in the range of strong interactions.

Developpements in this direction have to overcome the difficulties :

1) Why are there no similar hadron-antihadron resonances at lower energies ?
2) The $\Omega\bar{\Omega}$ -model predicts <u>an abundance of strange quark hadrons in the final state</u>, whereas experimentally π-mesons by far seem to dominate. However this defect was recently

removed by assuming that non-strange quarks are easier to be created or emitted than strange quarks.[104i,j]

3) Replacing in eq. (4.58) m_ψ = 3.095 Gev by $m_{\psi'}$ = 3.684 Gev the hadronic ψ'-decay width would increase by a factor 100. This discrepancy is due to the small mass difference : $(m_\psi - 3\cdot m_\phi)$ in the phase space factor.

4.2.3. Charmonium

The following section presents a dynamical model for the earlier discussed hypothesis that ψ is a charmed quark bound state .The consequences of such hypothesis in higher symmetry schemes have been discussed above in section 4.1.1. which is considered as complementary to the following one.

The unification of weak and electromagnetic interactions suggested the existence of a fourth quark c (spin $\frac{1}{2}$) - the carrier of a new quantum number "charm" - whose mass is considerably larger than those of the usual quarks $q \equiv (p,n,\lambda)$: $m_q < 1$ Gev, $m_c > 1$ Gev. The framework of non-abelian gauge theories[105] of such an attempt assumes that their interaction is described by massless gauge fields(coloured vector gluons) giving rise to strong forces at large distances. The hypothesis that ψ is a bound state of two charmed quarks then leads to the following picture[106a]: at energies beyond the $c\bar{c}$ - threshold the total hadronic e^+e^- annihilation cross section scales. This behaviour should also be found in the range above the $\lambda\bar{\lambda}$ -threshold and well below the $c\bar{c}$ -threshold. However, large narrow enhancements are expected just above and below the $c\bar{c}$ -threshold which are due to $c\bar{c}$ -bound states and resonances (Fig. 70). These bound states may then be described in complete analogy with the positronium in QED[106d] thus giving the name "Charmonium"[106a].

Using a Coulomb potential the following decay width for three gluon production (the minimum number consistent with colour and charge-parity conservation) with subsequent conversion to hadrons,was obtained (Fig. 71) :

$$\Gamma_h = |M_h|^2 \cdot |\psi(o)|^2 = \frac{2}{9\pi} (\pi^2-9) \frac{5}{18} \alpha_s^3 (\frac{4}{3} \alpha_s)^3 \cdot m_c \quad (4.60)$$

where $\alpha_s \equiv g^2/4\pi$ is the quark-gluon coupling constant, m_c the charmed quark mass and $\psi(o)$ the bound state wave function at the origin. Analogously, the lepton decay width is

$$\Gamma_\ell = |M_\ell|^2 \cdot |\psi(o)|^2 = \frac{2}{9} \alpha^2 (\frac{4}{3} \alpha_s)^2 \cdot m_c \quad (4.61)$$

with $\alpha \approx 1/137$. The ratio Γ_h/Γ_ℓ then is independent of $\psi(o)$ and permits determination of $\alpha_s \approx 0.26$. This along with $m_c \sim 1.5$ Gev implies

$$\Gamma_\ell \approx 0.8 \text{ kev} \qquad \Gamma_h \approx 20 \text{ kev} \quad (4.62)$$

which is of the right order of magnitude. The radius of Charmonium is

$$R_n = \frac{3}{2} \frac{n^2}{\alpha_s \cdot m_c} \sim \text{several fm} \quad (4.63)$$

These calculations do not necessarily have to assume a Coulomb potential. Inspired by gauge theory,potentials like

$$V(\vec{r}) = \frac{C_1}{|\vec{r}|} + C_2 |\vec{r}| \quad (4.64)$$

have been used to determine the mass spectrum via non-relativistic Schrödinger equation.

Note that the hadronic and leptonic ψ-widths increase with growing m_c (whereas $\Gamma_{\eta_c \to \gamma\gamma}$ decreases). Charmonium-like models involve the assumptions of a small decay region compared to the spatial extension of the state and the importance of binding effects. The second one is critical since the binding energy $2m_c - M_\psi$ of the $c\bar{c}$-meson is of strong origin whereas in positronium $2m_e - M = \frac{\alpha^2}{4} m_e$ is small [109ℓ].

Outside the resonance region, the hadron to μ-pair ratio is estimated as :

$$R \equiv \sigma_h/\sigma_\mu = \sum_q Q_i^2 \cdot (1 + \frac{\alpha_s}{\pi}) + \sum_c Q_i^2 \cdot \Theta(q^2 - 4m_c^2) \cdot F(q^2) \qquad (4.65)$$

The second term is due to charmed quark production ; its form is fixed by multi-gluon contributions in the $c\bar{c}$-blob (Fig. 72). There is a variety of immediate consequences and difficulties that such a type of approach has to explain [107]:

1) <u>There should exist a paracharmonium with $J^{PC} = 0^{-+}$</u> with a small mass difference : $\Delta M = M_{ortho} - M_{para} \approx 60$ Mev which is identified with η_c [73]. Its hadronic width is estimated to be considerably larger since it can decay by two-gluon emission : eg. $\Gamma_h(\text{para}) \sim 6\text{-}10$ Mev using

$$\Gamma_h(\text{para})/\Gamma_h(\text{ortho}) = \frac{5}{6} \cdot \frac{2}{9\pi} (\pi^2 - 9) \cdot \alpha_s \approx 0.013 \ . \qquad (4.66)$$

Two-photon emission is given by

$$\Gamma_{\eta_c \to 2\gamma} = \frac{4\pi}{3} \alpha^2 \frac{|\psi(o)|^2}{m_c^2} \cong 4 \text{ kev} \ . \qquad (4.67)$$

2) One expects <u>many more enhancements which are interpreted as bound states or resonances between (p,n,λ,c) quarks</u>, below and above ψ , due to radial and orbital excitations of charmonium-like systems. This latter possibility has been pursued using the non-relativistic Schrödinger equation and supposing a potential linearly increasing with distance r, to prevent escape of the quarks from each other, with the result that :

$$M_n = 2 \cdot m_c + \sqrt{\frac{K^2}{m_c}}\; a_n \qquad (4.68)$$

$(-a_n)$ is the nth zero of the Airy function Ai(x) on the negative real axis. Identifying $M_{1,2}$ with $\psi(3.1)$ and $\psi(3.7)$ one finds $m_c = 1.16$ Gev and $K = 0.211\ \text{Gev}^2$ (109).

3) Electromagnetic transition among these states should result in <u>numerous monochromatic photons with varied energies in the range of several tens to several hundreds of kev</u> . In particular, the modes : $\eta_c \to \gamma\gamma$ or $\psi' \to \gamma + \eta_c$ should be found with one photon monochromatic (Fig.62). The detection of such photons will be crucial to the verification of the charm scheme and the physics behind it (109,110). None is found so far.

4) A charm-anticharm threshold should exist not too far from ψ (around 4 Gev?) giving a considerable increase in the

total cross section. Apart from this effect, a virtual bound state is anticipated $\psi_{virtuel}$, influencing the behaviour at charm threshold as follows :

$$R_{threshold} = 1.33 \cdot \sqrt{1-\frac{4m_c^2}{q^2}} \left\{1+\frac{2m_c^2}{q^2}\right\} |\psi(o)|^2 \qquad (4.69)$$

The sudden increase due to the threshold factors is dampened down by the continuum wave function which is supposed to have a strong q^2-dependence (Fig.73)[111].

Although this picture has a number of attractive features there is so far no conclusive experimental evidence.

4.2.4. Zweig Selection Rule

In the preceeding sections, we have discussed possible dynamical explanations for the small ψ-widths. We present here the selection rule mechanism which is essentially based on quarks and higher symmetry. Despite this fact we think we can place it among the dynamical attempts and present it here.

Duality between resonance poles and Regge asymptotic behaviour has led to quark duality diagrams. Zweig's suppression rule[112] states that in the quark diagram of a given process, the two ends of a quark line must belong to distinct particles; put another way, quark lines from a given hadron do not annihilate each other. Any process not conforming to this rule is suppressed (Fig.74).

Typical examples of its application are $\phi \to 3\pi$, $\phi \to \pi\gamma$, $\phi \to \pi\rho$, $\pi N \to \phi N$, $f' \to \pi\pi$, etc. Empirical tests prove it correct in most cases[113]. However, its breaking is difficult to ascertain since there are alternatives : the considered hadron might be a mixture of quarks and not contain only one type.

The identification $\psi \equiv (c\bar{c})$ demands either the appearance of charmed hadrons in its decay, or no strong decays, or violation of the rule. The decay $\psi' \to \psi + \pi^+\pi^-$, forbidden by the rule,exists,which is a difficulty for this rule in this scheme; however it was estimated to be slightly suppressed in comparence to $\rho' \to \rho + \pi\pi$ (see section 4.2.1.) . Higher order contributions in the dual model give alternatives[114].

The electromagnetic decays eg. $\psi \to \eta_c + \gamma$ or $\psi' \to \gamma\gamma + \psi$ similarly have to resort to some suppression mechanism such as Zweig's rule in order to reduce their widths.

We mention here a recent experimental test of Zweig's selection rule in $pp \to \phi + X$[115]; it was found that there is no indication for its operation in ϕ-production by pp collisions at $\sqrt{s} = 6.8$ Gev.

4.2.5. Resonance Models, Duality

Within the framework of the dual resonance model it was hypothesized that ψ-particles lie on C=-1 partners of the daughter trajectories of the Pomeron ; thus they could be excited closed string or ring states with masses : $M_n^2 = 4n/\alpha'$, $\alpha' = 0.85$ (116). However the narrow width is not a dynamical consequence of this picture and can only be achieved by selection rules due to new quantum numbers. Another investigation, considering the dual string model as a vortex line of the Higgs model, leads to quantization of the Regge slope. It is conjectured that the ψ-particles lie on a trajectory with slope $\alpha'(n=2) = \frac{1}{4}\alpha'(n=1)$ where $\alpha'(1) = 0.9\ \mathrm{Gev}^{-2}$ is the ordinary Regge slope. The mass formula here is : $M_n^2 = 4.4.n + M_o^2$ (116d).

Conventional dynamical concepts such as unitarity, Reggeization and duality are sufficient to explain the small ψ-width once one does admit that ψ (imagined as a $c\bar{c}$ - bound state) virtually disintegrates via a $D\bar{D}$ - pair ; this in turn is supposed to annihilate into non-charmed hadrons (Fig. 75 (117)). The sum over all intermediate states is approximated by Regge pole exchange with the trajectory : $\alpha_\psi = \alpha(o) + \alpha' m^2$. The decay width is proportional to

$$\Gamma_\psi / \Gamma_{\text{conv.}} \approx \left(\frac{m_\psi^2}{s_o}\right)^{\alpha(o)-1} \qquad (4.70)$$

and demands $\alpha(0) \approx -3$, a value not too far from eq.(4.6,7). Leaving aside problems such as off-shell corrections of the $D\bar{D}$-intermediate states, positivity etc., this model also provides explanations for the decay widths of ψ' and ψ'' and the conventional hadronic resonances. Since $m > 2\,m_D$ is assumed, the 4.1 enhancement is expected to decay into ψ'' charmed mesons :

$$\Gamma\,(\psi'' \to D\bar{D}) \approx 80 \text{ Mev} \qquad (4.71)$$

As an interesting characteristic we mention : if $\psi' \to \psi + \pi\,\pi$ decay is via a single loop (Fig. 76) one should expect enhancements due to kinematics in some of the invariant mass distributions ; such behaviour was found to be a general characteristic for triangle loops with three particles created [117].

Another consequence of virtual two-body decay $D\bar{D}$ is its considerable influence below threshold[118]. Suppose that the created hadron system may be approximated by resonance states R_n (Fig. 77). Such a process is described by the vacuum polarization amplitude $\pi(q^2)$ whose imaginary part near threshold is fixed by kinematic threshold factors as follows :

$$\text{Im}\,\pi(q^2) \propto v^{3/2}\,, \quad v \equiv 1 - \frac{4m_D^2}{q^2} \qquad (4.72)$$

Note, $\sigma_{D\bar{D}}$ = const. $(\frac{1}{q^2})$. Im $\pi(q^2)$ approaches asymptotic scaling behaviour. A once-subtracted dispersion relation then fixes the real part

$$\text{Re}\ \pi\ (q^2) = \frac{2}{3\pi}\left[1 + 3\ v\{\ 1 - \sqrt{-v}\ \ \text{arctg}\ \sqrt{-v}\ \}\right] \quad (4.73)$$

whose contribution below threshold $q^2 \lesssim 4\ m_D^2$ shows a rapid increase (Fig.78). The widths for $(\psi,\psi',\psi'') \rightarrow R_n$: $\Gamma_{\psi \rightarrow R_n} \propto |\pi(q^2)|^2$ obviously are very small as long as ψ, ... etc. are below threshold and substantially increase with growing Im $\pi(q^2)$ above threshold. The decay width ratios

$$\begin{aligned} \Gamma_\psi : \Gamma_{\psi'} : \Gamma_{\psi''} &= 1 : 3 : 10 \quad \text{(threshold)} \\ &= 1 : 12 : 530 \quad \text{(threshold + resonance)} \end{aligned} \quad (4.74)$$

are considerably improved if the existence of a resonance just above threshold is assumed (second set of data)[118].

Another attempt[119] uses 'Sakurai-Duality' between the high energy limit of $R = \sigma_h/\sigma_\mu$ and the low energy resonances ρ, ω, ϕ (see III-3.4). R_∞ was determined by the parton model with the quarks of the corresponding resonance only and $R(q^2)$ at low q^2 was fixed by the resonance parametrization. The sum-rule

$$\int_{4m^2}^{S_{Max}} ds.\ R(s) \cong \int_{s_o}^{S_{Max}} ds.\ R_\infty\ (s) \quad (4.75)$$

connected the two parametrizations. Applied to the new narrow resonances, such a scheme suggests that the 4.1 enhancement is the first recurrence of ψ (3.1)[119].

4.2.6. Zero-Pole system

The subsequent suggestion was proposed just after the discovery of ψ. The more detailed experimental information now available doesn't leave much chance for its survival. We mention it here to exhibit the full spectrum of proposed explanations.

In this model ψ is viewed as a conventional hadron resonance whose width appears narrow due to the closeness of a resonance and a CDD-pole (zero in the amplitude)[120]. Inserting the resonance parametrization of $e^+e^- \to \mu^+\mu^-$:

$$A_\ell (s) = g \frac{\{s - (m - \frac{i\gamma}{2})2\}}{s - (M - \frac{i\Gamma}{2})^2} \qquad (4.76)$$

into partial wave unitarity

$$A_\ell - A_\ell^* = 2i \text{ Im } A_\ell = \frac{R_\ell}{\sqrt{q^2}} A_\ell^* A_\ell \qquad (4.77)$$

the decay width results :

$$2 M \Gamma_{tot} = \frac{R_\ell}{\sqrt{q^2}} g \quad \{(M^2 - m^2 + \frac{\gamma^2}{4}) + m^2\gamma^2\} \qquad (4.78)$$

If the elasticity stays unchanged and the factor in brackets is very small the width can be considerably reduced due to the closeness of the zero and pole explicity introduced in the amplitude of eq. (4.76).
Its cross section shape is similar to the (usual) resonance background interference pattern with a minimum at the lower end (for $m < M$) due to the numerator zero.

However there is an asymptotic constancy of the amplitude.
The difficulties here are :

1) The inelasticity $R_\ell \propto R \equiv \sigma_h/\sigma_\mu \approx 30$ at the resonance, thus forcing the zero and the pole to be extremely close.
2) Why are there no zero-pole effects observed at lower energies ?
3) Such assumption enforces a characteristic phase contour pattern which experimentally should be determinable [121].
4) All channels are equally suppressed in contradiction to the experiment.

However, from an S-matrix point of view, we do not see any reasons 'a priori' excluding a closeby zero-pole system or -as another alternative- multipoles [122].

4.2.7. New Interactions

The dynamical characteristics of the new resonances gave arguments for a conjectured new interaction strength. One approach [123] supposes the ψ-particles (composed of charmed quarks) to decay via a triplet of massive gauge bosons, aside from one photon exchange . The interaction Lagrangian is

$$\mathcal{L}_I = i\cdot f\cdot(\bar{N}\,\gamma_\mu\,\frac{\vec{\tau}}{2}\,N)\,\vec{B}^\mu \tag{4.79}$$

with N describing the four quarks by a pair of doublets (which, for simplicity, are assumed to obey SU(2) gauge group with exact color symmetry) and $\vec{B}_\mu$ a triplet of neutral gauge fields. These were thought to be the carriers of a new interaction strength : $G' = \eta \cdot G_F$, $\eta \approx 150$.

We mention in this line of reasoning the earlier discussed attempt to unify weak, electromagnetic and strong interactions which uses colored gauge bosons in a theory with charmed quarks [124,125] (see section 4.1.4). Instead of identifying ψ with a charm-anticharm system, it might be identified with gauge bosons (colored gluons). A distinct characteristic then is that : $\psi \to \gamma$ + hadrons contributes a significant fraction of all decay modes. Such picture does allow $\psi' \to \psi + \pi^+\pi^-$ decay. Only one of the two ψ's can be identified with an elementary gluon, the other is composite, and two more such objects are demanded within 50-100 Mev.

The idea of a new interaction strength received support from phenomenological VMD-based arguments [126]. Assuming that the usual current-field identification of VMD is valid for ψ too :

$$\mathcal{L}_I = e\, A_\mu \cdot j^\mu \;, \qquad j^\mu = (g_\rho \cdot \rho^\mu + \cdots + g_\psi \cdot \psi^\mu) \qquad (4.80)$$

the $\psi \to$ hadron coupling constant

$$g_\psi \underset{\sim}{\sim} (4\text{-}6)\ \frac{f_{\gamma\psi}}{m_\psi^2}\ ,\ \frac{f_{\gamma\psi}}{m_\psi^2} \approx\ 2.5\ \alpha \qquad (4.81)$$

is then bigger than the usual ones ; this fact has lead to the proposal of a new interaction strength .

We also present in this section attempts which investigate the consequences of interactions like ψ-h-ψ . Some of these approaches might more logically be placed in different sections ; however we prefer to present them here due to their common feature of ψ-pair production in hadronic interactions.

In section 4.1.3 we briefly indicated the assignment of a new quantum number t = ±1,0 to fields J_o, $\bar{J}_o$ whose linear combinations then are the observed ψ-particles. We here present the consequences(127) :

(i) In purely hadronic processes ψ-pair production is possible which manifests itself in an increase in the ψ/π ratio to considerably large values at high energies.

(ii) If, in addition, the ν-ψ-$\bar{\nu}$ coupling is supposed to exist, the total (νp) cross section would be much larger than the observed one. Conclusion : $|g_\nu| << |g_e|$ (Fig.79a). In deep inelastic scattering eN (resp.μN) - scattering ψ-production should be felt through the branching ratio at large q^2

$$\frac{ep \rightarrow e + \ell^+ \ell^- + h}{ep \rightarrow all} \sim 1 - 0.1\% \qquad (4.82)$$

due to the process drawn in Fig.79

(iii) If charged $\psi\pm$ exist, their couplings $e-\psi^{+}-\nu$ is estimated to be very small.

Further consequences of this picture are : ψ has isotopic spin and its charge is given by

$$Q = I_3 + \left(\frac{Y + t}{2} \right) \qquad (4.83)$$

The symmetry group for strong interactions in such case should have a higher rank than SU(3). CP-violating effects are expected in J_1-J_2 decay described by medium weak interactions of strength $\sim 10^{-6}$ (127).

There exist two further attempts in this direction which we will briefly present here. One of them (128) supposes that ψ-particles belong to a new class of heavy particles, called superons (S), which lie appreciably higher in masses than the familiar hadron multiplets. Their coupling strenghts are estimated to be

$$f_s \approx 0.01 \qquad (f_s : \text{h-S-h coupling})$$

$$\frac{g_s^2}{4\pi} \approx (0.1-1) \cdot \frac{g^2}{4\pi} \qquad (g_s : \text{h-S-S coupling})$$

$$\frac{h_s^2}{4\pi} \approx 20-70 \qquad (h_s : \text{S-S-S coupling})$$

$$\frac{g^2}{4\pi} \approx 5 \qquad (g : \text{h-h-h coupling}).$$

(4.84)

Predicted characteristics here are : superons which are sufficiently massive to decay into other superons should have decay widths which are anomalously large compared to what would be expected for hadrons. The decays of all the new heavy long-lived particles should be similar to ψ-decay.

The third attempt which supposes ψ-h-ψ couplings is in the framework of weak interactions. ψ and ψ' are identified with two neutral members of a U(3) triplet of W-bosons $W = (W^+, W^o, W^{o'})$ [129] :

$$\psi\,(3.1) \equiv W^{o'}$$
$$\psi'(3,7) \equiv W^o \qquad (4.85)$$

the charged one W^+ mediates weak interactions. Integer charge assignment requires to have a triality quantum number $t = 1$, $t_3 = \pm 1,0$ with

$$Q = I_3 + \frac{1}{2} Y + \frac{1}{3} t \equiv I_3 + \frac{1}{2} \bar{Y}. \qquad (4.86)$$

Quantum number assignment to W s is as in Table XX . If the triplet has strong interactions with hadrons (t = 0) it must be a strong pair type interaction. Weak interactions are assumed to violate triality.

The theory is based on a non-renormalizable Lagrangian

$$\mathcal{L}_I = \mathcal{L}_h + \mathcal{L}_W + \mathcal{L}_{em} \tag{4.87}$$

with

$$\mathcal{L}_h = \mathcal{L}_1(q) + \sum f \cdot J(W) \cdot F(q) + \sum f' \cdot J(W) \cdot J\ (W) \tag{4.88}$$

$$\mathcal{L}_W = g \cdot J_\mu \cdot W^\mu \tag{4.89}$$

where J(W) stands for the triality zero octet currents and F(q) represents the octet functions of quarks. The electromagnetic interaction $\mathcal{L}_{em}$ (which conserves triality) is fixed by the minimal replacement. Predicted consequences are :

(i) $\Gamma\ (W^+ \to \mu^\pm\ \nu) \sim 10$ kev. Weak non-leptonic decays of $W^\pm$ to hadrons are possible.

(ii) Decays ψ, $\psi' \to \ell^+\ell^-$ proceed by mechanisms as described in Fig.80 . The coupling strengths are anticipated in the order of $g \sim \frac{\alpha}{3}$

(iii) Proposed decay mechanisms for W^o, $W^{o'}$ $\rightarrow$ hadrons are drawn in Fig. 81 . The ratio

$$\Gamma(\psi \rightarrow \ell^+\ell^-) \Big/ \Gamma(\psi \rightarrow \text{hadrons}) \sim \left(\frac{1}{3f}\right)^2 \approx \frac{1}{20}$$

is of the right order of magnitude.

The decay $\psi' \rightarrow \psi + \pi\pi$ is considered to be strong.

(iv) For ψ photoproduction the right order of magnitude is obtained if the assumption of asymptotic "softening" of strong pair couplings is accepted.

(v) A crucial test will be the near equality of σ^ν and $\sigma^{\bar\nu}$ as well as $d\sigma^\nu/dy$ and $d\sigma^{\bar\nu}/dy$ (for lare y) in very high energy neutrino reactions(129) .

4.2.8. Varia

Having presented the important dynamical interpretations in the framework of hadrons and quarks, we now briefly discuss some related suggestions.

An interesting regularity was recently observed which is also respected by the new resonances ψ and ψ'. The particles life times τ_i, which are inverse proportional to their decay widths, obey the law

$$\tau_i = \tau_n \cdot \alpha^{x_i}, \tag{4.90}$$

τ_n is the neutron life time. Surprisingly the numbers x_i are integers apart from small deviations which appear to be systematic[130].

Measurement of the fraction of charged energy to total energy

$$x_i \equiv \Gamma_i^{ch} / \Gamma_i \tag{4.91}$$

of a particular decay channel i imposes constraints on specific models by the following <u>theorem</u>[131].

We first define the fractions of charged decays of particular decay channels. $x \equiv \Gamma^{Ch}_{tot} / \Gamma_{tot}$ is the charged fraction of all ψ-decays. $x_h \equiv \Gamma^{Ch}_h / \Gamma_h$ is the charged fraction of all purely hadronic decays. The charged fraction for purely leptonic decays is obviously $x_\ell = 1$. x_j ... abbreviates the set of charged fractions appearing in decay modes involving photons or lepton pairs in addition to hadrons.

The total fraction of charged particles

$$X \equiv \frac{\Gamma^{Ch}}{\Gamma_{tot}} = \frac{\Sigma x_i \cdot \Gamma_i}{\sum_i \Gamma_i} = \frac{\Gamma_e + \Gamma_\mu + \Gamma_h + \Sigma\, \Gamma_j}{\Gamma_e + \Gamma_\mu + x_h \cdot \Gamma_h + \Sigma x_j \cdot \Gamma_j} \qquad (4.92)$$

then is limited by the 'photonic' decays, $\psi \to \gamma$+hadrons etc. by the following relations:

$$F\,(\rho) \geqslant X \geqslant F\,(\sigma) \qquad (4.93)$$

where

$$F(y) = y \cdot \{1 - 2\frac{\Gamma_e}{\Gamma_{tot}} - \frac{\Gamma_h}{\Gamma_{tot}}\} + \{2 \cdot \frac{\Gamma_e}{\Gamma_{tot}} + x_h \cdot \frac{\Gamma_h}{\Gamma_{tot}}\} \qquad (4.94)$$

$\rho \equiv \text{Max}\{X_j\}$, $\sigma \equiv \text{Min}\{X_j\}$ are the maximal resp. minimal fraction X_j of charged hadron production in any mode involving photons and/or leptons. Unitarity might give further constraints.

4.3. Non-Hadronic Schemes

In this section we present suggestions which assume the ψ s to be non-strongly interacting particles.

4.3.1. Weak Neutral Vector Bosons

The unification of weak and electromagnetic interactions led to the postulation of a neutral weak current with the Hamiltonian

$$H = \sum_{\ell \equiv e,\mu} j_\mu\,(\ell) \,.\, Z_o^\mu \qquad (4.95)$$

$$j_\mu\,(\ell) = \bar{u}_\ell \left\{ g_V^\ell + g_A^\ell\, \gamma_5 \right\} \gamma_\mu\, u_\ell$$

Its characteristics are a vector and axial vector coupling (where one of two still might be absent). Suppose that ψ is identified with the neutral weak boson Z_o responsible for the neutral current : $\psi \equiv Z_o$[132].

The experimental values for the leptonic decay widths

$$\Gamma_e = \frac{g_V^2+g_A^2}{12\pi} \cdot m_\psi \sim 5 \text{ Kev} \Longrightarrow \frac{g_V^2+g_A^2}{4\pi} \sim 5\cdot 10^{-6} \quad (4.96)$$

demand that the leptonic coupling constants are of the order of the Fermi coupling[65a] :

$$\frac{g_V^2+g_A^2}{m_\psi^2} \sim 6\cdot 10^{-6} \text{ Gev}^{-2} \Longrightarrow \lambda \frac{G_F}{\sqrt{2}} \quad , \quad \lambda \cong 0.8 \quad (4.97)$$

(the usual Fermi coupling is $G_F/\sqrt{2} = 7\cdot 10^{-6} \cdot m_p^{-2}$). Note that the ordinary vector mesons such as ρ, ω, ϕ... etc. have leptonic widths of the same order of magnitude.

On the basis of lepton-hadron universality one would expect universal couplings of leptons and quarks to Z_o. The hadronic couplings although of the same order of magnitude, are not the same as the leptonic ones. Estimates in specific models predict ratios of hadronic to leptonic coupling constants to be

$$g^h/g^\ell \gtrsim 1.5^{(65b)} \text{ resp. } \gtrsim 4^{(65b)} . \quad (4.98)$$

ψ would therefore not be universally coupled.

One furthermore expects parity violating effects in $e^+e^- \to \mu^+\mu^-$ (unless there is pure vector or axial vector coupling) which show up in a modified angular distribution (compared to the one of QED : $d\sigma/d\Omega = \frac{\alpha 2}{4q^2}\{1 + \cos^2\theta\}$) and consequently a charge-asymmetry A (θ) in the polarization P_μ (θ) of the final state μ^- and in a modified inclusive angular distribution $d\sigma/d\cos$.
Experimentally there is no indication of such effects. The inclusive distribution $d\sigma/dx$ would presumably be unchanged(133).

Such an interpretation has a number of further consequences - some of them in disagreement with the experimental results(134,65) :

1) In earlier investigations the Z_o - mass was estimated to be $M_{Z_o} > 37$ Gev and deep inelastic neutrino experiments indicate similar orders of magnitude ($m > 4$ Gev). The total cross sections σ ($\nu(\bar{\nu})$ N) bend off much earlier for smaller Z_o - masses.

   If $M_Z o \sim 3$-5 Gev this would require a substantial variation in the inclusive neutral to charged current ratio due to the Z_o propagator

$$d\sigma\ (\nu + N \to \mu^- + X) \,/\, d\sigma\ (\nu + N \to \nu + X) \approx \left(\frac{1}{q^2 - M_\psi^2}\right)^2 \qquad (4.99)$$

   which has not been so far. The estimated ratio of hadronic to leptonic coupling constants are in disagreeement with e^+e^- estimations.

In Fermi theory the total cross sections for $\nu_\mu e \to \nu_\mu e$ or $\bar{\nu}_\mu e \to \bar{\nu}_\mu e$ scattering increase linearly for asymptotic energies (violating unitarity) ; due to a Z_o-propagator effect they would reach a constant value. Experimental results in the present energy range do not permit detection of the anticipated propagator effects.

2) If the Z_o mass is as small as 3-4 Gev, then why are there no effects for the charged $W^\pm$ - bosons ?

3) How is the existence of two ψ's explained - are there several gauge bosons ? In gauge theories unifying weak and electromagnetic interactions, the universal gauge coupling constants g are necessarily larger than the electric charge e. In theories with several intermediate bosons, then $g_i > e$. Thus, if ψ, ψ',... were gauge bosons, their couplings to leptons and hadrons would be at least as large as e, implying leptonic widths many times larger than are observed. Artificial alternatives are still possible[65a]. By giving up unifying attempts and just maintaining renormalizability of weak interactions, one gains a freedom in the choice of the coupling constants. Neutral and charged current phenomena are uncorrelated.

If renormalizability is no longer imposed, we are back to conventional weak interaction theory (with all its difficulties) and we might interpret $\psi \equiv W$ as a weak boson, with the difficulties mentioned above.

5) Within the quark parton model : $\Gamma\gamma\to h / \Gamma\gamma\to e^+e^- = R = \sum^N Q_i^2$. If similar ideas apply to $\psi \equiv Z_o$ (neutral weak boson) then $\Gamma\psi\to h / \Gamma\psi\to e^+e^- = N$. Experimentally this ratio is $\sim$ 14 at ψ (3095) and $\gtrsim$ 100 at ψ (3685)[65a].

6) The decay $\psi' \to \psi + \pi^+\pi^-$ was estimated to be slightly suppressed,but still stronger than weak interactions . In addition,if $\psi \equiv Z_o$ there should be comparable lepton and π-rates in $\psi' \to \psi + X$[65a]. One might allow strong coupling for ψ and ψ' when they couple in pairs. This would enhance the total cross section for ψ- production in deep inelastic ep and νp scattering : σ (ep $\to$ e+ ψ + X)[65] (see section 4.2.7.) .

7) ψ has been generated in photoproduction $\gamma N \to \psi N$ with a rate which is compatible with, and indicates it to be, a strongly interacting particle.

A variety of modified forms of such attempts lead to similar characteristics[135,134]. Other suggestions introduce either two types of spin-1 gauge bosons, thus replacing the Cabbibo rotation by their mixing[137], or identify ψ

and ψ' with the two neutral members of a U(3) triplet of W-bosons(129) : (W^+, W^o, $W^{o'}$). They are supposed to carry triality quantum number t = 1 and enjoy a strong pair interaction with hadrons. The small width $\Gamma_e \cong 5$ kev of $\psi(3.1)$ is explained by the inverse of the diagram in Fig. 80 . $\psi(3.1)$ should then be produceable in pairs via strong interactions with threshold at 32.2 Gev. The production cross section for both narrow resonances at NAL energies are predicted to be large. W^+ is expected around 4 Gev (see section 4.2.7.).

One might enlarge the Weinberg-Salam group $SU_2 \times SU_1$ to

$$SU_{1B} \times SU_{1M} \times SU_{1E}$$

and identify the narrow resonances with the gauge bosons which are connected with the baryonic, mesonic and electric charges(138). This model has a large number of independent parameters and e-μ universality is spoiled by the decay of ψ (3.7). In spite of its clumsiness it stands as a representative that neutral currents are not yet out. Another attempt uses $SU_2 \times U_1 \times U_1$(139). Higgs particle interpretation can be excluded due to spin(140).

4.3.2. Heavy Photon

Given the situation that ψ and γ are dynamically identical objects one then might reconsider theories of indefinite metric[141] - a possibility which perhaps has been given up to early (?). In the electromagnetic interaction Lagrangian

$$T = e.J_{\mu} \cdot (A^{\mu} + i\ B^{\mu}). \qquad (4.100)$$

the usual zero-mass positive metric photon field A_{μ} is replaced by a complex field. B_{μ} is the negative metric heavy photon field of mass m_B.

At present such attempt is in an unsatisfactory state because :

1) The partial width of B^o decaying to the lepton modes is

$$\Gamma\ (B^o \rightarrow e^+e^-) = \frac{2}{3}\,\alpha \cdot m_B \approx 15 \text{ Mev} \qquad (4.101)$$

a value much to large compared with the data. Further development might perhaps lead to a more realistic value.

2) There are sofar no indications of QED-violations in $e^+e^-\rightarrow\mu^+\mu^-$ or $e^+e^-\rightarrow e^+e^-$ off the narrow resonances and the lower limits of the cut-off masses are estimated : $\Lambda >$ 20 Gev[1,7] (see section 3.2.3.) .

3) If the cross section for $e^+e^-\rightarrow e^+e^-$ would show interference effects, destructive before and constructive after the peak, such pattern could not be explained by a resonance ansatz (additional to the usual direct

and crossed channel photon poles, as e.g. intermediate vector boson) and rather points in the direction of indefinite metric theories [142]. Experimentally there clearly are interference effects in both e^+e^- and $\mu^+\mu^-$ final state channels (Fig.17) .

4) One can argue against the existence of negative norm intermediate states which this type of theories assume. Unitarity is satisfied if the existence of complex photon poles on the physical sheet are assumed which however lead to difficulties with causality .

The characteristics of this model are : it introduces an additional pole in the photon propagator which does not modify the dynamics, apart from an increase in the cross section close to m_B^2, and there is an asymptotic fall off like q^{-4}.

V. Conclusion

This paper grew out of the desire to understand e^+e^- -annihilation- its experimental results as well as the proposed theoretical schemes. It naturelly splits up in three parts : presentation of the experimental results, modelling on $e^+e^- \to$ hadrons and theoretical explanations of ψ.

In the second part of this paper we learned about the new results on the total and inclusive cross sections in e^+e^- annihilation which show no smooth scaling behaviour as expected from single photon QED investigations, but instead exhibit a resonance structure and a rising ration $R \equiv \sigma_h/\sigma_\mu$ Two of these resonances are extremely narrow , whereas a further one is broad. The normalized inclusive distributions change considerably in the large x-region on and off the narrow resonances.

In the third part we presented the scaling arguments. Our main attempt here was to find a common descriptive basis of all proposed models and to draw general conclusions.

There exists a wealth of proposed explanations, investigations and speculations on the ψ-particles which we try to cover in the <u>fourth part</u>. Our main aim here is to classify and to group the enormous amount of theoretical ideas. Although many of these schemes are quite attractive we can not escape the impression that we are still far from the correct explanation.

Acknowledgements

I would like to thank J.Ellis for his reading ,critics and suggestions of several versions of this paper. Discussions with F.Close,G.Schierholz,M.Chaichian,F.Schrempp, P.Grassberger helped to progess in the understanding.

I am grateful to Prof. M.Guenin for his encouragements.

LIST OF REFERENCES

(1)	B. Richter	London Intern. Conf. (1974) p. IV-37 "$e^+e^- \to$ Hadrons".
(1a)	R.R.Larsen	SLAC-PUB-1479 (Sept.1974) "e^+e^- reactions-Experiments"
(2)		See Refs. (9, 10, 11, 12)
(3a)	F.J. Gilman	London Intern. Conf. (1974) "Deep inelastic scattering and the structure of hadrons".
(3b)	R. Gatto	Aix-en-Provence, Intern. Conf. (1973) "Electromagnetic interactions of hadrons".
(3c)	G. Altarelli	Riv. del NC. 4, 3 (1974) 335 "The physics of deep inelastic phenomena".
(4a)	J. Ellis	London Intern. Conf. (1974) "Theor. ideas about $e^+e^- \to$ hadrons at high energies".
(4b)	O.W. Greenberg	Maryland U. (Oct. 1974) "e^+e^- reactions - theory".
(4c)	J.D. Bjorken B.L. Ioffe	SLAC-PUB-1467 (Aug. 1974) "Annihilation of e^+e^- into hadrons".
(4d)	C.H. Llewellyn-Smith	Erice Lectures, (1974) "Is theoretical physics able to explain e^+e^- annihilation into hadrons".
(4e)	B. Humpert	Helv. Phys. Acta 47, 4 (1974) 491 "Models for the annihilation of $e^+e^- \to$ hadrons".
(5)	F.J. Gilman	SLAC-PUB-1537 (Febr. 1975) "Electron-psitron annihilation and the structure of hadrons".
(6)	H.Lynch	See Ref. (10g)
(7a)	B.L. Beron et al.	Phys. Rev. Lett. 33, 11 (1974) 663 "Observation of the reactions $e^+e^- \to e^+e^-$, $e^+e^- \to \gamma\gamma$ and $e^+e^- \to \mu^+\mu^-$".
(7b)	J.E. Augustin et al.	Phys. Rev. Lett. 34, 4 (1975) 233 "Measurement of $e^+e^- \to e^+e^-$ and $e^+e^- \to \mu^+\mu^-$".
(8)	G. Grammer, Jr. J. Smith	Phys. Rev. D11, 1 (1975) 223 "Multiplicity measurements in e^+e^- colliding beam experiments".

BROOKHAVEN

(9a) J.J. Aubert et al. — Phys. Rev. Lett. 33, 23 (1974) 1404
"Experimental observation of a heavy particle J".

(9b) J.J. Aubert et al. — Phys. Rev. Lett. 33, 27 (1974) 1624
"Non-observation of heavier J particles from p-nucleon reactions".

(9c) J.J. Aubert et al. — MIT (Jan. 1975)
"Discovery of the new particle J".

(9d) U.Becker — MIT,Cambridge (May 1975)
"Massive,narrow resonances".

SLAC-SPEAR

(10a) J.E. Augustin et al. — Phys. Rev. Lett. 33, 23 (1974) 1406
"Discovery of a narrow resonance in e^+e^- annihilation".

(10b) G.S. Abrams et al. — Phys. Rev. Lett. 33, 24 (1974) 1453
"The discovery of a second narrow resonance in e^+e^- annihilation".

(10c) R.L. Ford et al. — Phys. Rev. Lett. 34, 10 (1975) 604
"Measurements of $e^+e^- \rightarrow e^+e^-$, $e^+e^- \rightarrow \mu^+\mu^-$ and $e^+e^- \rightarrow \gamma\gamma$ at CM-energy close to 3105 Mev".

(10d) A.M. Boyarski et al. — Phys. Rev. Lett. 34, 12 (1975) 762
"Search for narrow resonances in e^+e^- annihilation in the mass region 3.2 to 5.9 Gev".

(10e) J.E. Augustin et al. — Phys. Rev. Lett. 34, 12 (1975) 764
"Total cross section for hadron production by electron-positron annihilation between 2.4 Gev and 5.0 Gev CM-energy".

(10f) G.S. Abrams et al. — Phys. Rev. Lett. 34, 18 (1975) 1181
"The decay of ψ (3684) into ψ (3095)".

(10g) H. Lynch — SLAC-PUB-1536 (Febr. 1975)
"Recent results for e^+e^- annihilation at SPEAR".

(10h) J.A. Kadyk et al. — Lawrence Berkeley Lab. (April 1975)
"Some properties of the ψ(3.7) resonance and features of the total hadronic cross section e^+e^- annihilation from 2.4 Gev to 5.0 Gev CM energy".

(10i) G.J. Feldman, M.L. Perl — Physics Report (May 1975)
"Electron-positron annihilation above 2 Gev and the new particles".

(10k) A.M. Boyarski et al. — Phys. Rev. Lett. 34, 21 (1975) 1357
"Quantum numbers and decay widths of the ψ (3095)".

(10ℓ)	A.M.Boyarski et al.	Lawrence Berkeley Lab. (May 1975) "Limits on charmed meson production in e^+e^- annihilation at 4.8 Gev CM-energy".
(10m)	M.L.Perl	SLAC (June 1975) "Lectures on electron-positron annihilation - Part II".
(10n)	V.Lüth	SLAC-PUB-1599 (June 1975) "Quantum numbers and decay modes of the resonances ψ(3095) and ψ(3684) ". (Palermo Conf.1975)
(10o)	G.Feldman	Palermo Conf. 1975,Talk
(10p)	C.Morehouse	1975 Spring Meeting of the American Physical Society.

ADONE

(11a)	C. Bacci et al.	Phys. Rev. Lett. 33, 23 (1974) 1408 "Preliminary results of Frascati (Adone) on the nature of a new 3.1 Gev particle produced in e^+e^- annihilation" E : Phys. Rev. Lett. 33, 27 (1974) 1649.
(11b)	W.W. Ash et al.	Lett. al. NC. 11, 17 (1974) 705 "Experimental study of the new 3.1 Gev particle e^+e^- collision at Adone".
(11c)	R. Baldini-Celio et al.	Lett. al NC. 11, 17 (1974) 711 "Experimental results on the production and decay modes of the 3101 Mev Resonance at Adone".
(11d)	G. Barbellini et al.	Lett. al. NC. 11, 17 (1974) 718 "Preliminary results on the energy dependence of the production of collinear relativistic particles at Adone in the 3.1 Gev CM-energy region".
(11e)	C. Bacci et al.	Lett. al. NC. 12, 8 (1975) 269 "Experimental results on the reaction $e^+e^- \rightarrow$ photons at the 3.1 Gev resonance".
(11f)	B. Bartoli et al.	Frascati LNF-74/64 (Dec. 1974) "Observation of a possible anomaly in the μ-pair angular distribution at total CM-energies around s = 3.1 Gev".

DESY-DASP

(12a)	W. Braunschweig et al.	Phys. Lett. B 53, 5, (1975) 393 "A measurement of large angle e^+e^- scattering at the 3100 Mev resonance".
(12b)	L. Criegee et al.	Phys. Lett. B 53, 5 (1975) 489 "Confirmation of the new 3700 Mev resonance in e^+e^- collisions".
(12c)	W. Braunschweig et al.	Phys. Lett. B 53, 5 (1975) 491 "Measurement of collinear and nearly collinear photon pairs produced by e^+e^- annihilation at the 3100 Mev resonance".
(12d)	W. Braunschweig et al.	Phys. Lett. B 56, 5 (1975) 491 "Muon pair production by e^+e^- annihilation at the 3100 Mev resonance".
(12e)	B. Wiik	CERN-seminar, (Febr. 1975).
(12f)	W.Braunschweig et al.	DESY-DASP 75/14 (May 1975) "Two-body hadronic decays of the 3.1 Gev resonance".
(13)	CERN	Theory Notice Board : Preliminary Informations.

HADRONIC PRODUCTION

(14a)	J.J.Aubert	See Ref. (9a)
(14b)	J. Ellis	See Ref.
(14c)	J.J.Aubert	See Ref. (9b)
(14d)	CERN Workshop	See Ref.
(14e)	P.H. Frampton V. Rabl	Syracuse U. "Consequences of a direct ψ-lepton coupling".
(14f)	B. Knapp	Phys. Rev. Lett. 34, 16 (1975) 1044 "Di-muon production by neutrons".
(14g)	F.W.Büsser et al.	Phys.Lett. B 56, 5 (1975) 482 "Observation of high mass electron-positron pairs produced in proton-proton collisions at the CERN ISR".

PHOTOPRODUCTION

(15a)	D.E. Andrews et al.	Phys. Rev. Lett. 34, 4 (1975) 231 "Search for photoproduction of ψ (3105)".
(15b)	D.E.Andrews et al.	Phys.Rev.Lett. 34, 17 (1975) 1134 "An improved upper limit for photoproduction of ψ(3105) near shreshold".
(15c)	J.F.Martin et al.	Phys.Rev.Lett. 34,5 (1975) 288 "Experimental upper limit on the photo-production cross section for ψ(3105)".
(15d)	J.T.Dakin et al.	Phys.Lett. B 56,4 (1975) 405 "Muon pair photoproduction at 20.5 Gev".
(15e)	U.Camerini et al.	Phys.Rev.Lett. 35,8 (1975) 483 "Photoproduction of the ψ-particles".
(15f)	B.Knapp et al.	Phys.Rev.Lett. 34, 16 (1975) 1040 "Photoproduction of narrow resonances".
(15g)	H.J.Berend et al.	Phys.Lett. B 56, 4 (1975) 408 "Photoproduction of ϕ-mesons at small t-values".

SCALING

(16)	T.D. Lee	CERN 73-15, Lab. I (1973) "Scaling properties and the bound state model of physical baryons".
(17)	S.D. Drell D.J. Levy T.M. Yan	Phys. Rev. D 1, (1972) 1617 "Theory of deep inelastic lepton nucleon scattering and lepton-pair annihilation processes".
(18)	J.D. Bjorken	Phys. Rev. 148, 4 (1966) 1467 "Application of the chiral U (6) x U (6) algebra of current densities".
(19)	D.J. Fox et al.	Phys. Rev. Lett. 33, 25 (1974) 1504 "Early tests of scale invariance in high energy muon scattering"
(20a)		Experiment by the European μ-Beam Collaboration at CERN.
(20b)	N.S. Craigie G. Schierholz	Ref. Th. 2014-CERN (May 1975) "On extracting the deep inelastic structure of Reggeons from high energy muon-proton experiments".
(20c)	J.D.Sullivan	Phys. Rev. D 5, 7 (1972) 1732 "One pion exchange and deep inelastic electron-nucleon scattering".
(21a)	V.N. Gribov L.N. Lipatov	Phys. Lett. B 37, (1971) 78 "Deep inelastic electron scattering in perturbation theory".
(21b)	V.N. Gribov L.N. Lipatov	Sov. J. of Nucl. Phys. 15 (1972) 438 "Deep inelastic ep scattering in perturbation theory".
(21c)	V.N. Gribov L.N. Lipatov	Sov. J. of Nucl. Phys. 15 (1972) 675 "e^+e^- pair annihilation and deep inelastic ep scattering in perturbation theory".
(22a)	P.M. Fishbane J.J. Sullivan	Phys. Rev. D 6, (1972) 3568 "Inelastic e^+e^- annihilation in perturbation theory".
(22b)	P.M. Fishbane J.J. Sullivan	Phys. Rev. D 7, (1973) 1879 "Scaling behaviour of the cut-off vector-gluon model".

(22c)	P.M. Fishbane	Lett. al. NC. B 53, 1 (1974) 111 "Reciprocal relation in asymptotically free theories".
(23a)	Y. Eylon Y. Zarmi	Nucl. Phys. B 83, (1974) 475 "On the hadron-parton reciprocity relation".
(23b)	W.S. Lam Y. Zarmi	Bielefeld U. Bi-74/19 (Sept. 1974) "On scaling terms in e^+e^- annihilation".
(23c)	Y. Zarmi	Phys. Lett. B 53, 4 (1974) 351 "Simple tests of quark-nucleon reciprocity".
(24)	A. Di Giacomo K.I. Konishi	Phys. Lett. B 51, 1 (1974) 67 "Scaling in e^+e^- annihilation into hadrons".
(25a)	G. Schierholz M.G. Schmidt	Phys. Lett. B 52, 4 (1974) 467 "Dual light-cone model predictions for total and inclusive e^+e^- annihilation".
(25b)	G. Schierholz M.G. Schmidt	Ref. Th. 2013 - CERN (April 1975) "Total and inclusive e^+e^- annihilation 'understood' from a scaling model for deep inelastic electron scattering".
(26a)	F.E. Close W.N. Cottingham	Ref. Th. 2009 - CERN (April 1975) "The kinematics and symmetry structure of $e^+e^-\rightarrow$ hadron pairs".
(26b)	S.J. Brodsky G.R. Farrar	SLAC-PUB-1473 (Aug. 1974) CALT-68-441 "Scaling laws for large momentum transfer processes".
(27a)	J.D. Bjorken S.D. Drell	"Relativistic Quantum Mechanics" McGraw-Hill Book Company, London.
(27b)	J.Werle	Warsaw Univ., IFT/74/14 (Oct. 1974) "The physical vacuum as a quantum liquid".

Introductory papers are :

(28a)	F. Close	Daresbury Lecture Notes No. 12 (1973) "Partons and Quarks".
(28b)	K. Huang	MIT (Aug. 1972) "Partons".

(28c)	V. Barger R. Phillips	Wisconsin "Quark parton model relations in deep inelastic lepton scattering".
(28d)	G. Altarelli	Ref. (3c).

PHENOMENOLOGICAL PARTON MODEL

(29a)	S. Ferrara M. Greco A.F. Grillo	Lett. al. NC. 4 (1970) 1 "e^+e^- annihilation into hadrons".
(29b)	N. Cabibbo G. Parisi M. Testa	Lett. al. NC. 4, 1 (1970) 35 "Hadron production in e^+e^- collisions".
(29c)	S. Berman J.D. Bjorken J. Kogut	Phys. Rev. D 4, 11 (1971) 3388 "Inclusive processes at high transverse momentum".
(29d)	G. Preparata R. Gatto	Nucl. Phys. B 67, (1973) 362 "One-particle and two particle inclusive deep-inelastic electron-positron annihilation in a massive quark model".
(29e)	G. Preparata	Phys. Rev. D 7, 10 (1973) 2973 "Massive quarks and deep-inelastic phenomena".
(29f)	R. Gatto G. Preparata	Riv. del NC. 4, 4 (1974) 445 "Theor. Studies for high energy e^+e^- collisions".
(29h)	R. Gatto G. Preparata	Phys. Lett. B 50, 4 (1974) 479 "Senile scaling in e^+e^- annihilation".
(29i)	M.B. Einhorn G.C. Fox	NAL-PUB 74/75 THY (June 1974) "Precocious scaling and duality in the quark-parton model - a reformulation".
(29k)	N.S. Craigie A.B. Kraemmer K.D. Rothe	NC. A 11, 3 (1972) 645 "Model amplitudes for e^+e^- annihilation into pions at high energies".
(29l)	F.M. Renard	Montpellier U. (Febr. 1972) "Hadronic structure of the photon : Partons versus vector mesons".
(29m)	F. M. Renard	Phys. Lett. B 40, 4 (1972) 484 "Are there oszillations in $e^+e^- \rightarrow$ hadrons total cross section ?".

(29n) H. Inagaki — Tokyo Metropolian Univ. (1975)
"Correlated parton model and e^+e^- reaction".

COVARIANT PARTON MODEL

(30a) P.V. Landshoff, J.C. Polkinghorne, R.D. Short (LPS) — Nucl. Phys. B 28 (1971) 225
"A non perturbative parton model of current interactions".

(30b) P.V. Landshoff, J.C. Polkinghorne — Physics Reports C 5 (1972) 1
"Models for hadronic and leptonic processes at high energy".

(30c) P.V. Landshoff, J.C. Polkinghorne — Phys. Rev. D 6, 12 (1972) 3708
"Connection between electroproduction and annihilation".

(30d) R.L. Kingsley, P.V. Landshoff, C. Nash, J.C. Polkinghorne — Nucl. Phys. B 65 (1973) 397
"Multiplicities in electroproduction and e^+e^- annihilation".

(30e) H. Osborn, G. Woo — Nucl. Phys. B 62 (1973) 413
"Approach to asymptopia in deep-inelastic scattering and annihilation".

(30f) J.C. Polkinghorne — Phys. Lett. B 49, 3 (1974) 277
"Quark-quark interactions and high-energy physics".

FIELD THEOR. PARTON MODELS

(31a) S.D. Drell, D.J. Levy, T.M. Yan — Phys. Rev. D 1 (1972) 1617
"Theory of deep-inelastic lepton nucleon scattering and lepton-pair annihilation processes. III. Deep-inelastic e^+e^- annihilation".

(31b) I.A. Sanda — Phys. Rev. D 8, 12 (1972) 4510
"A field theoretic formulation of the parton model".

(31c) P.M. Fishbane, J.D. Sullivan — Phys. Rev. D 6, 12 (1973) 3568
"Inelastic e^+e^- annihilation in perturbation theory".

(31d)	G. Altarelli L. Maiani	Nucl. Phys. B 51 (1973) 509 "Deep-inelastic processes in a ladder model".
(31e)	N.S. Craigie A.B. Kraemmer K.D. Rothe	NC. A 18, 2 (1973) 188 "Predictions of quark-parton model for e^+e^- annihilation into hadrons".
(31f)	H. Cheng T.T. Wu	FERMILAB-PUB (August 1974) "Remarks on e^+e^- annihilation into hadrons in quantum field theory".
(31g)	A.M. Polyakov	Sov. Phys. JETP 33 (1971) 850 "Similarity hypothesis in strong interactions. II. Cascade production of hadrons and their energy distribution associated with e^+e^- annihilation".

BAG-MODELS

(32a)	A. Chodos R.L. Jaffe K. Johnson C.B. Thorn V.F. Weisskopf	Phys. Rev. D 9, 12 (1974) 3471 "A new extended model of hadrons".
(32b)	W.A. Bardeen M.S. Chanowitz S.D. Drell M. Weinstein T.M. Yan	SLAC-PUB-1490 (Sept. 1974) "Heavy quarks and strong binding : a field theory of hadrons structure".
(32c)	P. Vinciarelli	Phys. Lett. B 53 (1975) 457 "Genesis of field theory bags and e^+e^- annihilation".
(33)	B. Humpert	Geneva U. (in preparation) "Approaches in e^+e^- annihilation".

PARTONS WITH STRUCTURE

(34a)	M.S. Chanowitz S.D. Drell	Phys. Rev. Lett. 30 (1973) 807 "Speculations on the breakdown of scaling at 10^{-15} cm"
(34b)	M.S. Chanowitz S.D. Drell	Phys. Rev. D 9 (1974) 2078 "Speculation on the breakdown of scaling at 10^{-15} cm".

(34c)	M. Pavkovic	Phys. Lett. B 46, 3 (1973) 435 "The breakdown of scaling due to the possible structure of nucleonic constituents".
(34d)	G.B. West	Phys. Rev. D 10, 1 (1974) 329 "Possible evidence for quark substructure from electron-positron colliding beam experiments".
(34e)	G.B. West P. Zerwas	Phys. Rev. D 10, 7 (1974) 2130 "Experimental consequences of quark-structure".
(34f)	M. Böhm H. Joos M. Krammer	DESY 74/4 "Electromagnetic properties of hadrons in a relativistic quark model".
(34g)	F. Close	Rutherford Lab. (1973) (W. Toner + R. Zia, Ed.) "Relations between weak and electromagnetic interactions".

SPIN-1 PARTONS

(35a)	J. Cleymans G.J. Komen	Nucl. Phys. B 78, 3 (1974) 396 "Charged spin one partons ?".
(35b)	M.A. Furman G.J. Komen	Nucl. Phys. B 84, 2 (1975) 323 "Scaling behaviour and charged spin one partons in a gauge model".
(35c)	S.C. Matinyan S.V. Esaibegyan	JETP Lett. 19, 6 (1974) 227 "Vector partons ?".
(35d)	G. Venturi S. Zerbini	Bologna U. (1974) "Another model for the violation of scaling".

VECTOR MESONS

Introductory reviews are :

(36a)	M. Greco	Erice Lecture Notes (1974) "Deep inelastic process from a non-orthodox point of view".
(36b)	F. Renard	Nucl. Phys. B 82, 1 (1974) 1 "Vector mesons and e^+e^- annihilation".
(36c)	D. Schildknecht	Erice Lecture Notes (1974) "Is vector meson dominance still alive ?".

(37) M. Greco — Nucl. Phys. B 63 (1973) 398
"Deep-inelastic processes".

(38a) J.J. Sakurai — Phys. Lett. B 46, 2 (1973) 207
"Duality in $e^+e^- \rightarrow$ hadrons ?".

(38b) F.M. Renard — Phys. Lett. 493, 5 (1974) 477
"About a possible constancy of the total cross section of the e^+e^- annihilation into hadrons".

(38c) G.J. Gounaris — Nucl. Phys. B 68 (1974) 574
"Duality in the current propagator".

(38d) D. Schildknecht — DESY 74/39 (Aug. 1974)
"Comments on generalized vector dominance".

(39a) C.A. Dominguez, M. Greco — Lett. al NC. 12, 12 (1975) 439
"Charm, EVMD and narrow resonances in e^+e^- annihilation".

(39b) D. Schildknecht, F. Steiner — Phys. Lett. B 56, 1 (1975) 36
"New hadronic degrees of freedom, e^+e^- annihilation and deep inelastic scattering".

(40) B. Humpert — under investigation

(41) A. Bramon, E. Etim, S. Ferrara, M. Greco, A.F. Grillo, G.C. Rossi, Y. Srivastava — Frascati - LNF 73/68 (Dec. 1973)
"Theoretical aspects of high energy electron-positron collisions".

(42) H. Sato — Lett. al NC. 12, 2 (1975) 43
"A new finite energy sum rule in e^+e^- annihilation."

(43) M. Chaichian, J. Cleymans, R. Peschanski — CERN Ref. Th. 1990 (March 1975)
"Semi-local duality in e^+e^- annihilation into hadrons".

FERMI MODEL

Introductory reviews to stat. theories are :

(44a) E.L. Feinberg — Physics Report C 5 (1972) 237
"Multiple production of hadrons at cosmic ray energies (experimental results and theor. concepts)".

(44b) H. Satz — Cracow Lectures, Zakopane, 1973 "The stat. description of multihadron production processes".

(44c) P. Carruthers — Ann. of the N.Y. Academy of Sciences 229, (1974) 91 "Heretical Models of particles production".

(45a) E. Fermi — Progr. Theor. Phys. 5 (1970) 570 "High energy nuclear events".

(45b) E. Fermi — Phys. Rev. 81 (1951) 683 "Angular distribution of the pions produced in high energy nuclear collisions".

(46) J. Engels, K. Schilling, H. Satz — NC. A 17, 4 (1973) 535 "Statistical and Thermodynamic descriptions of hadron production in e^+e^- annihilation".

(47) W.S. Lam, F. Suhonen — Phys. Lett. B 50, 4 (1974) 453 "Phase space and multihadron production in annihilation processes at high energies".

(48) H. Satz — CERN Ref. Th. 1905 (Aug. 1974) "Dynamical features in cluster models for multihadron production".

(49) N.N.Bogoliubov, D.V.Shirkov — "Introduction to the theory of quantized fields". Interscience Publ.,Inc. ,N.Y. 1959

LANDAU MODEL

(50a) D. Ter Haar — "Collected papers of L.D. Landau" Gordon and Breach, New-York, N.Y. (1965)

(50b) L.D. Landau, S.Z. Belenkij — NC. Suppl. 3, (1956) 15

(51a) F. Cooper, G. Fry, E. Schonberg — Phys. Rev. Lett. 32, 15 (1974) 862 "Electron-positron annihilation into hadrons and Landau's hydrodynamical model".

(51b) F. Cooper, G. Fry — Phys. Rev. D 11, 1 (1975) 192 "Landau's hydrodynamical model of particle production and electron-positron annihilation into hadrons".

(51c) F. Cooper — Los Alamos (Sept. 1974) "Landau's hydrodynamical model of particle production".

(51d) P.D. Morley — Nucl. Phys. B 85, 2 (1975) 471 "Hydrodynamical model of electron-positron annihilation into hadrons".

(51e) C.B.Chiu, E.C.G.Sudarshan, Kuo-Hsiang Wang — Texas Univ. (1975) "Hydrodynamical expansion with frame independence symmetry in high energy multiparticle production".

(51f)	E.V. Shuryak	Phys. Lett. B 34, 6 (1971) 509 "Final state interaction in high energy e^+e^- annihilation into hadrons".
(51g)	J. Baacke	Phys. Lett. B 49, 3 (1974) 297 "Predictions of the hydrodynamical model for e^+e^- annihilation into hadrons".
(51h)	G. Grammer, Jr. H.T. Nieh Y.P. Yao	SUNY, Stony Brook (June 1974) "Electron-positron annihilation : statistical and thermodynamic considerations".
(52a)	E.I. Feinberg	Phys. Lett. B 52, 3 (1974) 203 "Hydrodynamical approach to the electron-positron annihilation into hadrons".
(52b)	T.F. Hoang	Argonne (1975) "Viscosity effects in Landau's hydrodynamical model".

HEISENBERG MODEL

(53a)	W. Heisenberg	Z.f. Physik 133 (1952) 65 "Mesonerzeugung als Stosswellen Problem".
(53b)	W. Heisenberg	"Kosmische Strahlung" Berlin, 1953, p. 148.
(53c)	H. R. Rechenberg D.C. Robertson	Acta Phys. Polonica B 5 (1974) 507. "A semiclassical model of particle production".
(54)	P.B. Bart	Phys. Rev. Lett. 32, 19 (1974) 1080 "Solitary waves in non-linear field theories".

THERMODYN.MODEL

(55)	R. Hagedorn	CERN lectures 71/12 (may 1971) "Thermodynamics of strong interactions".
(56a)	J. Engels H. Satz K. Schilling	Phys. Lett. B 49, 2 (1974) 171 "The energy dependence of single particle spectra in e^+e^- annihilation".
(56b)	H.J. Möhring J. Kripfganz E.M. Ilgenfritz J. Ranft	Nucl. Phys. B 85 (1975) 221 "Multihadron production in nucleon-anti-nucleon annihilation at rest and e^+e^- annihilation in the stat. bootstrap model.

(56c)	H.J. Möhring	Nucl. Phys. B 87, 3 (1975) 509 "Multihadron producation in nucleon-anti-nucleon annihilation at rest and e^+e^- annihilation in the stat. boostrap model. II. Momentum distribution, average energies and transverse momenta".
(57a)	N.S. Craigie T.F. Walsh	DESY-74/56 (Nov. 1974) "Lessons on e^+e^- annihilation from simple chain emission models". (and references therein).
(57b)	A.L. Maison	Phys. Rev. D 10, 8 (1974) 2588 "Some models for the decay of time-like, highly virtual particles".

JET MODEL

(58)	R. Baier J. Engels H. Satz K. Schilling	Bielefeld Univ. (Jan. 1975) "Jet structure in e^+e^- annihilation".

CASCADE DECAY

(59a)	S.J. Orfanidis V. Rittenberg	Phys. Rev. Lett. 33, 7 (1974) 455 "Gaussian distributions in electro-production and e^+e^- annihilation".
(59b)	S.J. Orfanidis	Phys. Rev. 10, 9 (1974) 2892 "Connection between branching processes and scale invariant field theory for e^+e^- annihilation".
(59c)	V. Rittenberg D.H. Schiller	Nucl. Phys. B 87, 3 (1975) 480 "Do we see anomalous dimensions in e^+e^- inclusive annihilation at Q = 4 Gev ?".
(59d)	A.M. Polyakov	Sov. Phys. JETP 32, (1971) 296 "A similarity hypothesis in the strong interactions I. Multiple hadron production in e^+e^- annihilation".
(59e)	A.M. Polykov	Sov. Phys. JETP 33, (1971) 850 "Similarity hypothesis in strong interactions II. Cascade production of hadrons and their energy distribution associated with e^+e^- annihilation".
(59f)	P. Olesen H.C. Tze	Niels Bohr Inst. (1974) "Multiplicity scaling in e^+e^- annihilation and Polyakov's hypothesis".

NEW INTERACTIONS

(60a)	O.W. Greenberg G.B.C. Yodh	Phys. Rev. Lett. 32, 25 (1974) 43 "Diffractive lepton scattering and constant $\sigma(e^+e^- \to h)$; a new regime in lepton physics".
(60b)	D.V. Nanopoulos S.D.P. Vlassopoulos	Lett. al. NC. 10, 17 (1974) 751 "Implications of the recent '$e^+e^- \to$ hadrons + X' data on the leptonic world".
(60c)	R. Chanda	Lett. al NC. 11, 13 (1974) 593 "Leptons with hadron core ?".
(61a)	I. Bigi J.D. Bjorken	Phys. Rev. D 10, 11 (1974) 3697 "Direct coupling of hadrons to leptons".
(61b)	H. Schlereth	Lett. al NC. 11, 6 (1974) 337 "Some remarks on electron-quark interactions".
(61c)	M.A. Bég G. Feinberg	Phys. Rev. Lett. 33 (1974) 606 "Exotic interactions of charged leptons".
(61d)	O.W.Greenberg A.Raychaudhuri	Maryland U. (1975) "$d\sigma/d\Omega$ for $e^+e^- \to hX$ from direct quark-lepton interactions".
(62a)	J.C. Pati	Phys. Rev. Lett. 32, 19 (1974) 1083 "Are there anomalous lepton-hadron interactions ?".
(62b)	J.D. Davies J.G. Guy R.K.P. Zia	NC. A 24, 3 (1974) 324 "Experimental limits to the e^+e^- rate and their consequences for a direct electron-hadron coupling".
(62c)	A. Soni	Phys. Lett. B 52, 3 (1974) 332 "The structure of anomalous lepton hadron interactions and $\pi^0 \to e^+e^-$".
(62d)	A. Soni	Phys. Lett. B 53, 3 (1974) 280 "Implications of anomalous lepton hadron interactions".
(62e)	T.S Romanova I.S. Satsonkevich	Sov. J. of Nucl. Phys. 18, 2 (1974) 198 "On the $\pi^0 \to e^+e^-$ decay".
(62f)	A.B. Lahanas	Durham Univ. (Sept. 1974) "A Consistent approach to $e^+e^- \to h + X$ and deep inelastic scattering".
(63a)	T. Goldman P. Vinciarelli	Phys. Rev. Lett. 33, 4 (1974) 246 "Is the cross section at SPEAR time dependent ?".
(63b)	F.Ma	UC, Santa Barbara "$e^+e^- \to$ hadrons : The rôle of the Higgs scalar".

(63c)	D.W. McKay H. Munczek	Phys. Rev. D 11, 3 (1975) 688 "e^+e^- annihilation into hadrons and bridging Higgs particles".
(64a)	D. Fryberger	SLAC-PUB-1535 (Febr. 1975) "On possible implications of a large σ ($e^+e^- \to$ hadrons)".
(64b)	S. Wolfram	EtonCollege, Windsor (March 1975) "A hadronic core model of leptons".
(64c)	S. Wolfram	Eton College, Windsor (April 1975) "Hadronic electrons ?".

CHARM

(65a)	I. Bars et al.	SLAC-PUB-1522 (Jan. 1975) "Survey of theor. speculations on the nature of ψ (3105) and ψ (3695)".
(65b)	Theory Group CERN	CERN-Int. Report (Dec. 1974) "The narrow peaks in pp $\to e^+e^-$ + X and e^+e^- annihilation".
(65c)	H. Harari	SLAC Internal Notes (Nov. 1974) "ψ-chology".
(65d)	M. Fukugita S. Yazaki (Workshop)	Tokyo U. (Jan. 1975) "Proc. of the discussion meeting on the new narrow resonances".
(65e)	J. Ellis	Ref. Th 1996 - CERN (March 1975) "e^+e^- annihilation, the new particles and charm".
(66)	M.K. Gaillard B.W. Lee J.L. Rosner	Rev. Mod. Phys. 47, 2 (1975) 277 "Search for charm".
(67a)	J.D. Bjorken S.L. Glashow	Phys. Lett. 11, 3 (1964) 255 "Elementary particles and SU(4)".
(67b)	S.L. Glashow J. Iliopoulos L. Maiani	Phys. Rev. D2, (1970) 1285 Weak interactions with lepton-hadron symmetry".
(68)	J. Iliopoulos	London Intern. Conf. (1974) "Gauge theories".
(69)	J.L. Rosner	Physics Reports 11C, 4 (1974) p. "The classification and decays of resonant particles".

(70,71,72) We list here the <u>papers suggesting SU_4</u> (not mentioned otherwise in this report).

(70a) S. Borchardt, V.S. Mathur, S. Okubo
Phys. Rev. Lett. <u>34</u>, 4 (1975) 236
"SU(4) symmetry and the possible existence of new hadrons".

(70b) S. Borchardt, V.S. Mathur, S. Okubo
Rochester University (Nov. 1974)
"A possible explanation of the new resonance in e^+e^- annihilation ".

(70c) S. Eliezer, B.R. Holenstein
Massachusetts University (Dec. 1974)
"ψ(3105) and SU(8) symmetry".

(70d) B.G. Kenny, D.C. Peaslee, L.J. Tassic
Phys. Rev. Lett. <u>34</u>, 7 (1975) 429
"Quark-model comparison on the narrow e^+e^- resonances".

(70e) C.S. Kalman
Concordia Univ., Montreal (April 1975)
"Are there four ψ particles ?".

(70f) D.H. Boal, R.H. Graham, J.W. Moffat, P.J. O'Donnell
Phys. Rev. Lett. <u>34</u>, 9 (1975) 541
"SU(4) explanation of the narrow resonances in e^+e^- annihilation."

(70g) D.H. Boal, R.H. Graham, J.W. Moffat, P.J. O. Donald
Toronto Univ. (Dec. 1974)
"An SU(4) explanation of the narrow resonances in e^+e^- annihilation in a model without strageness - changing currents".

(70h) J.W. Moffat
Toronto Univ., Canada (Nov. 1974)
"Some consequences concerning the new-resonance discovered in e^+e^- annihilation".

(70i) Z. Maki, I. Umemura
Kyoto Univ. (Dec. 1974)
"A subnuclear approach to ψ' particles".

(70j) Z. Maki, K. Ohnishi, T. Teshima, I. Umemura
Kyoto Univ. (Febr. 1975)
"The 'ψ' particles in the quartet model".

(70k) M. Inose, R. Sugano, T. Tsujimura
Osaka City Univ. (Dec. 1974)
"An interpretation of resonances ψ(3105) and ψ(3695) in terms of quartet model based on SU(4) > SU(2) x SU(2)".

(70ℓ) T. Tsujimura
Osaka City Univ. (March 1975)
"On decays of new resonance ψ's, and dominance of the peripheral interactions".

(71a)	C. Itoh I. Minamikava K. Miura T. Watanabe	Meiki-Gakuin Univ. (Dec. 1974) "Charmed mesons".
(71b)	M. Koshiba S. Matsuda	Tokyo Univ. - HEP (Dec. 1974) "Charmed particles and their decays".
(71c)	M. Nakagawa	Meijo Univ. Nagoya (March 1975) "New particles in the framework of quartet model".
(71d)	Y. Hara	Tokyo Univ. of Education (Jan. 1975) "The ψ-particles and the quartet model".
(71e)	C. Montonen M. Roos N. Tornqvist	Lett. al NC. 12, 16 (1975) 627 "Meson mass relations and mixing in U(4) x U(4)".
(71f)	R.V. Caffarelli	Phys. lett. 55 B, 5 (1975) 481 "Meson decouplets in the SU(4) x SU(4) scheme. A possible explanation of narrow e^+e^- resonance".
(71g)	P. Marcolungo	Padova Univ. (Dec. 1974) "The 3105 Mev meson and SU(4) breaking".
(71h)	D.B. Lichtenberg	Indiana Univ. (Jan. 1975) "Hadron mass formulas in a charmed quark model".
(71i)	Fayyazuddin Ryazuddin	Islamabad Univ, Pakistan (Febr. 1975) "Decay width of resonance ψ in broken SU(4)".
(71j)	S. Iwao	Kanazawa Univ. (Febr. 1975) "A comment on Schwinger-like and $c\bar{c}$ bound-state models of ψ-particels".
(71k)	F. Ravndal	Nordita, Copenhagen (April 1975) "Orthocharmonium and $\psi \to H\bar{H}$".
(71ℓ)	S. Rudaz	Cornell Laboratory (Dec. 1974) "Decay modes of ψ(3695)".
(71m)	G.J. Aubrecht M.S.K. Razmi	Oregon Univ. (March 1975) "Some meson coupling constants in broken SU(4) and the radiative decays of ϕ_c and η_c".

(72a)	Y.Iwasaki	Phys.Rev. Lett. 34 , 22 (1975) 1407 "SU(4) symmetry and nonleptonic decays".
(72b)	J.Ellis M.K.Gaillard D.V.Nanopoulos	Ref.Th. CERN (May 1975) "On the weak decays of high mass hadrons".
(72c)	P.J.O'Donnell C.L.Ong	Toronto U. (May 1975) "The quark model and SU(4)".
(72d)	R.Vergara- Caffarelli	Pisa U. (May 1975) "Baryon resonances and SU(4)"
(72e)	M.Matsuda M.Tanimoto	Hiroshima U. (May 1975) "Universality of the quark-orbit excitation mass level and the breaking of SU(4) symmetry".
(72f)	A.Kazi G.Kramer D.H.Schiller	DESY 75-10 (May 1975) "Mass formulas for broken SU(4)".
(72g)	D.H.Boal R.H.Graham J.W.Moffat	Toronto U. (May 1975) "SU(4) symmetry and the decays of the new hadrons".
(72h)	J.W.Moffat	Toronto U. (May 1975) "SU(4) and SU(8) mass spliting and the possible existence of the new hadrons".
(72i)	D.H.Boal R.H.Graham J.W.Moffat P.J.O'Donnell	Toronto U. (May 1975) "SU(4) symmetry and the decays of the ψ and ψ'".
(72j)	J.W.Moffat	Toronto U. (May 1975) "Charmed-particle production in $\bar{\nu}$ interactions and the SU(4) quark model".

PSEUDOSCALAR η_c

(73a) B.W. Lee, C. Quigg — FERMILAB-74/110 (Dec. 1974) "Some considerations on η_c".

(73b) R.F. Dashen, I.J. Muzinich, B.W. Lee, C. Quigg — FERMILAB-75/18 (Jan. 1975) "Coherent production and decay modes of a pseudoscalar partner of the ψ(3105) boson".

(73c) I. Bars et al. — see Ref. 65 a

(73d) R. Blankenbecler et al. — SLAC-PUB-1531 (Jan. 1975) "SLAC workshop notes on the narrow states in hadronic and photonic experiments".

(73e) C.G. Callan et al — Phys. Rev. Lett. 34, 1 (1975) 52 "Remarks on the new resonances at 3.1 and 3.7 Gev".

(73f) M.K. Gaillard, B.W. Lee, J.L. Rosner — FERMILAB-PUB 75/74 (Jan. 1975) "Addendum : Search for charm".

(73g) T. Appelquist, H.D. Politzer — Phys. Rev. Lett. 34, 1 (1975) 43 "Heavy quarks and e^+e^- annihilation".

(74a) S. Kitakado, S. Oritio, T.F. Walsh — DESY 74/75 (Nov. 1975) "Remarks on new meson states".

(74b) S. Kitakado, S. Oritio, T.F. Walsh — DESY (Jan. 1975) "The width of ψ (3105)".

SPECTRAL FUNCTION SUM RULES

(75a) M.P. Khanna — Imperial College (March 1975) "Spectral function sum rules, asymptotic SU_4 and vector meson leptonic decays".

(75b) T. Hagiwara, R.N. Mohapatra — City College, N.Y. (Jan. 1975) "Comment on the interpretation of J-particles as a charm-anticharm bound state".

(75c) D.A. Dicus — Phys. Rev. Lett. 34, 13 (1975) 846 "Leptonic decays of vector mesons in SU(4)."

(75d) F.J. de Urries, J. Léon, A. Tiemblo — Madrid - Rocasolano (April 1975) "Electromagnetic gauge invariance hadron interactions and the new particles."

(75e)	Z. Maki K. Ohnishi T. Teshima I. Unemura	Kyoto Univ. (Febr. 1975) "The 'ψ' particles in the quartet model."

ALGEBRAIC APPROACH

(76a)	E. Takasugi S. Oneda	Phys. Rev. Lett. 34, 15 (1975) 988 "Asymptotic SU(4) in the e^+e^- annihilation of new résonances". E : PRL 34, 18 (1975) 120
(76b)	E. Takasugi S. Oneda	Phys. Lett. 34, 17 (1975) 1129 "Purely algebraic approach to the new narrow resonances."
(76c)	E. Takasugi S. Oneda	Maryland Univ. (March 1975) "Quark charge assignment and the ρ, ω, ϕ and ψ (3105) $\rightarrow e^+e^-$ decays".
(76d)	E. Takasugi S. Oneda	Maryland Univ. (Febr. 1975) "New narrow boson resonances and SU(4) symmetry selection rules, SU(4) mixing and mass formulas".

(77)	T. Goto V.S. Mathur	Rochester Univ. (Febr. 1975) "An alternative quark model with charm."
(78a)	R.D. Field C. Quigg	Fermilab - 75/15 THY (Jan. 1975) "Estimates of associated charm production cross sections".
(78b)	M.B.Einhorn C.Quigg	Fermilab-Pub-75/21 - THY (February 1975) "Nonleptonic decays of charmed mesons: Implications for e^+e^- annihilation".

COLOUR

(79)	M. Gell-Mann	Phys. Lett. 8 (1964) 214 A schematic model of baryons and mesons."
(80a)	H. Fritzsch M. Gell-Mann	Proc. of NAL Conf. (1972) 132 "Current algebra; quarks and what else ?"
(80b)	J. Schwinger	UCLA (Febr. 1975) "Speculations concerning the ψ particles and dyons."
(81a)	M. Han Y. Nambu	Phys. Rev. 139 (1965) 1006 "Three triplet model with SU(3) symmetry."
(81b)	H. Lipkin	Phys. Rev. D 7 (1973 1850 "Clashing symmetries in unified descriptions of electromagnetic and weak interactions and the case of the Han-Nambu model."
(82a)	O.W. Greeberg	Maryland Univ. (Jan. 1975) "Electron-positron annihilation into hadrons and color symmetries of elementary particles or the out look for color : gray or rosy ?"
(82b)	I. Bars et al.	see Ref.65 a .

(83) We list here the papers suggesting $SU_3 x\ SU_3^c$ (not mentioned otherwise in this report).

(83a)	F. Close J. Weyers	Louvain Univ. (March 1975) "Charmless colorful models of the new mesons."
(83b)	F. Close	Lecture Notes (1975) "How high are higher symmetries and ψ - chology for J-walkers."
(83c)	M. Arik D.D. Coon S. Yu	Brookhaven, BNL (March 1975) "Color Nonet."
(83d)	M. Kuroda Y. Yamaguchi	Tokyo Univ. (April 1975) "The color-scheme for new resonances and its comparison with the charm-scheme."

(83e)	S.B. Gerasimov A.B. Govorkov	JINR, Dubna (1975) "Interpretation of ψ-mesons in the three triplet model with integral charges."
(83f)	Fayyazuddin Riazuddin	Islamabad Univ. (March 1975) "Possible explanation of the new resonances ψ and ψ' in e^+e^- annihilation in Nambu-Han model."
(83g)	N. Marinescu B. Stech	Heidelberg Univ. (March 1975) "The new mesons and broken $SU_3' \times SU_3''$ (colour)."
(83h)	B. Stech	Heidelberg Univ. (Dec. 1974) "Partons, currents and the new mesons."
(83i)	B.Stech	Heidelberg U. (Febr. 1975) "Broken Color symmetry"
(83j)	S. Hori T. Suzuki A. Wakasa E. Yamada	Kanazawa Univ. (Nov. 1974) "A comment on the classification of new resonances ψ or J ."
(83k)	M. Koca Z.Z. Aydin	Lett.al NC. 13,4 (1975) 153 "Does recently discovered particle ψ (3105) imply Han-Nambu model ?"
(83ℓ)	S.Y. Tsai	Nikon Univ. (Dec. 1974) "Interpretation of the ψ particles in the framework of the three-triplet model."
(83m)	W. Alles	Lett. al Nuovo Cim. 12, 9 (1975) 285 "On the Han-Nambu model and ψ (3105) and ψ (3695) mesons."
(83n)	D. Bailin D.R.T. Jones A. Love	Lett. al Nuovo Cim. 12, 10 (1975) 375 "Remarks concerning the e^+e^- resonance at 3.1 GeV."
(83 o)	S.G.Kenny D.C.Peasley L.J.Tassie	Phys.Rev.Lett. 34 , 23 (1975) 1482 "Modified Han-Nambu model for the e^+e^- resonances".
(83p)	B.G.Kenny D.C.Peaslee L.J.Tassie	Australian Nat.Univ. (May 1975) "Meson masses in the modified Han-Nambu model".
(83q)	T.C. Yang	Phys. Lett. 56 B, 2 (1975) 161 "A model of coloured vector mesons."
(83r)	H. Suura M. Kuroda	Tokyo Univ. (Dec. 1974) "Role of singlet filters in the coloured quartet model of hadrons."

(83s)	D.B. Lichtenberg	Indiana Univ. (Dec. 1974) "Colour and Charm in particle physics."
(84a)	A.I. Sanda H. Terazawa	Rockefeller Univ. (March 1975) "Does a fourth ψ particle exist at about 4.9. GeV ?"
(84b)	G.Feldman P.T.Matthews	Imperial College (ICTP - 74/12) "Has SLAC already seen ψ_4 ?".
(84c)	G.Feldman P.T.Matthews	ICTP,Trieste (May 1975) "A case for color".
(84d)	G.Feldman P.T.Matthews	Imperial College (ICTP - 74/16) "Coloured vector mesons and the sum rule for complete Okubo mixing".
(85a)	I. Bars R.D. Peccei	Phys. Rev. Lett. 34, 15 (1975) 985 "The colourful ψ's. "
(85b)	M. Krammer D. Schildknecht F. Steiner	DESY (Dec. 1974) "J(3.1), ψ(3.7) - How about Colour ? "
(85c)	D.Schildknecht	DESY 75/13 (May 1975) "Color and the new particles. A brief review".
(85d)	S. Kitakado T.F. Walsh	Lett. al Nuovo Cim. 12, 14 (1975) 547 "Colour versus charm."
(86)	S.B. Gerasimov A.B. Govorkov	Dubna (Dec. 1974) "Generalized Weinberg sum rules and width of decay of new neutral vector mesons into lepton - antilepton pairs."
(87)	M. Suzuki	Phys. Lett. 56 B, 2 (1975) 165 "Fourth triplet quark for ψ ."
	OTHER IDEAS	
(88)		See Refs. 127 - 129
(89a)	T.N. Pham B. Pire T.N. Truong	Paris, Ecole Polytechnique (Dec. 1974) "Consequences of a new quantum number for the 3165 MeV and 3695 MeV resonance."
(89b)	S. Minami	Osaka City University (Dec. 1974) "Note on the narrow resonance with $M \cong 3.1$ GeV. "
(89c)	S. Minami	Osaka City University (Dec. 1974) "Narrow resonance ψ's, a new quantum number and strength of interaction."

(89d)	S. Minami	Osaka City University "Remarks on the new resonance ψ's ."
(90a)	T. Das P.P. Divakaran L.K. Pandit V. Singh	Phys. Rev. Lett. 34, 12 (1975) 770 "ψ-particles, SU_4 and anomalous currents."
(90b)	T. Das P.P. Divakaram L.K. Pandit V. Singh	Tata Inst. (Jan. 1975) "A test of models for the $\psi(3.1)$ using the branching ratio into charged particles."
(90c)	T. Das P.P. Divakaran L.K. Pandit V. Singh	Pramana 4, 3(1975) 105 "The ψ-particles in an SU_4 scheme with anomalous currents."
(90d)	A. Khare	Phys. Rev. Lett. 34, 22 (1975) 1410 "Quark anomalous magnetic moment and ψ particles."
(91)	F. Wilczek	Princeton Univ. "Possible degeneracy of heavy quarks."
(92a)	R.M. Barnett	Phys. Rev. Lett. 34, 1 (1975) 41 "Model with three charmed quarks."
(92b)	R.M. Barnett	Harvard Univ. (Febr. 1975) "A charmed quark model for narrow resonances."

(93a)	H. Harari	SLAC - PUB 1568 (March 1975) "A new quark model for hadrons."
(93b)	H.Harari	SLAC-PUB-1589 (May 1975) "An analysis of a new quark model of hadrons".
(94)	A. Soni	Phys. Rev. Lett. 34, 6 (1975) 352 "Is the photon g-spin blind ? "
(95)	T.S. Santhanam B. Gruber	Würzberg Univ. (April 1975) "Are the J-particles members of a G(2) multiplet ?"
(96)	M.Hongoh	Montréal U. (Mathematics Dept.) (May 1975) "Dynamical SU(3) model for strong interactions and ψ particles "

LEPTON QUARKS

(97a)	G.L. Godfrey	LBL - Berkeley (Nov. 1974) "Lepton quarks"
(97b)	E.J.Konopinski H.M.Mahmoud	Phys.Rev. 92 (1953) 1045 "The universal Fermi interaction".
(98a)	O.W. Greenberg C.A. Nelson	Phys. Rev. D 10, 8 (1974) 2567 "Composite models of leptons."
(98b)	O.W. Greenberg C.A. Nelson	Maryland Univ. 74-006 (July 1973) "A schematic model of leptons and its possible relations to hadrons."
(99a)	S.Minami	See Refs . 89 b - d
(99b)	S. Minami	Lett. al Nuovo Cim. 12, 5 (1975) 129 "Some considerations about e^+e^- ($\bar{p}p$) interactions, muon-electron universality and strangeness scheme for leptons."
(100)	J.C.Pati A.Salam	See Refs. 124

INTERFERENCE EFFECTS

(101a)	J.D. Jackson	LRL-Berkeley (Dec. 1974) "Radiative corrections and resonance parameters in e^+e^- annihilation."
(101b)	K. Geer G.B. Mainland W.F. Palmer S.S. Pinsky	Ohio State Univ. (Dec. 1974) "Electromagnetic production and decay of ψ (3105) ."

(101c) F.A. Berends, K.J.F. Gaemers, G.F. Komen — Leiden Univ. (Jan. 1975) "Resonances and interference effects in scattering and μ-pair production."

(101d) B. Humpert — Lett. al Nuovo Cim. 12, 13 (1975) 473 "Does the "3105" cause vacuum polarisation effects ?"

(101e) S. Iwao, M. Shako — Lett. al Nuovo Cim. 12, 14 (1975) 513 "Phenomenological approach to ψ(3.1) particles."

(101f) Ngee-Pong Chang — City College,N.Y. (May 1975) "Implications of photon mixing for hadronic decay of ψ(3095) : Interference and new width parameters".

(101g) J. Bjorken et al. — SLAC-PUB-1515 (Dec. 1974) "Notes from the SLAC theory workshop on the ψ ."

PHOTOPRODUCTION

(102a) T.L. Neff, C. Sachrajda, D. Sivers, J. Townsend — SLAC-PUB-1540 (Febr. 1975) "An ω-mixing model for the production of non-hadronic ψ's ."

(102b) C.E. Carlson, P.G.O. Freund — Enrico Fermi Inst. (Dec. 1974) "Suppressed ψ-photoproduction : a test for the charm hypothesis."

(102c) T. Goldman — SLAC-PUB-1538 (Febr. 1975) "Upper bound on charm quarks in nucleons from the cross section for ψ (3100) photoproduction."

(102d) R. Aviv, Y. Goren, D. Horn, S. Nussinov — Tel Aviv Univ. (April 1975) "Suppression of ψ-photoproduction."

(102e) M.K. Gaillard, B.W. Lee, J.L. Rosner — Rev. Mod. Phys. 47, 2 (1975) 277 "Search for charm."

(102f) R. Blankenbecler et al. — see Ref. 73 d

(102g) R. Brower, J. Primack — Santa Cruz Univ. (Dec. 1974) "Why is ψ-photoproduction small ? "

(102h)	V. Barger R.J.N. Phillips	Wisconsin Univ. (Febr. 1975) "Implications of ψ N scattering of universal features of other elastic interactions."
(102i)	G.J. Gounaris A. Verganelakis	Lett.al NC. 13,7 (1975) 241 "On diffractive photoproduction and electroproduction of the ψ's ."
(102j)		See Refs.73d
(102k)	T. Inami	Phys. Lett. 56 B, 3 (1975) 291 "A broken SU(4) scheme for pomeron couplings and the diffractive photo- and neutrino production of vector mesons."
(102ℓ)	B.L.Young K.E.Lassila	Iowa State U. (May 1975) "Photoproduction of J(ψ) particles and predictions for proton Compton scattering".

$\psi' \rightarrow \psi + \pi^+\pi^-$

(103a)	J.D. Jackson	LRL-Berkeley (Dec. 1974) "On the decay ψ (3695) $\rightarrow$ ψ (3105) + $\pi\pi$."
(103b)	B.J. Harrington S.Y. Park A. Yildiz	Harvard Univ. (June 1975) "Scalar meson (ε) dominance model in the decay $\psi' \rightarrow \psi + 2\pi$."
(103c)	J. Schwinger K.A. Milton W.Y. Tsai L. DeRaad, Jr.	UCLA (April 1975) "Resonance interpretation of ψ' (3.7) into ψ (3.1) ."
(103d)	W.Y. Tsai L. DeRaad, Jr. K.A. Milton	UCLA (May 1975) "Resonance model description of the decay $\psi(3.1) \rightarrow \pi^+\pi^- + \gamma$."
(103e)	F.M. Renard	Lett.al NC. 13,7 (1975) 247 "Remarks about $\psi' \rightarrow \psi + 2\pi$ and $\psi \rightarrow \omega + 2\pi$."
(103f)	R. Decker M. Moreno	Univ. de Louvain (April 1975) "An analysis of $e^+e^- \rightarrow \psi' \rightarrow \psi\pi\pi$ and the ε resonance."
(103g)	L.S. Brown R.N. Cahn	Phys.Rev.Lett. 35,1(1975) 1 "Chiral symmetry and $\psi' \rightarrow \psi\pi\pi$ decay."
(103h)	R.N.Cahn	Fermilab-Pub-75/46 (June 1975) "On the separation of $\psi \rightarrow \pi\pi+\gamma$ from $\psi \rightarrow \pi^+\pi^-\pi^0$ ".
(103i)	D. Morgan M.R. Pennington	Rutherford Lab. (April 1975) "The $\psi' \rightarrow \psi + \pi\pi$ decay as a test of PCAC."

(103j)	H.B. Thacker	Stony Brook (Dec. 1974) "Is ψ (3695) $\rightarrow$ ψ (3105) + $\pi\pi$ a strong decay ? "
(103k)	S. Iwao	Kanzawa University (Dec. 1974) "A model of ψ particles and a mechanism of their interaction."
(103ℓ)	K.I. Matumoto	Toyama University "A comment on the decay mode ψ(3700) $\rightarrow$ ψ (3100) + $\pi^+\pi^-$."
(103m)	J. Pasupathy	Tata Inst. (Febr. 1975) "Hadronic interpretation of narrow resonances at 3.105 and 3.695 BeV."
(103n)	H. Primakoff L. Pilachowski W.A. Simmons S.F. Tuan	Hawaii Univ. (May 1975) "On the relation between the decays J(3.1) $\rightarrow \pi^+ + \pi^- + \gamma$ and ψ(3.7) $\rightarrow \pi^+ + \pi^-$ + J(3.1) ."
(103o)	H.Goldberg	Northeastern U. (April 1975) "Color gluons and the decay of the ψ(3700) into ψ(3100)".
(103p)	G.J.Feldman F.J.Gilman	SLAC-PUB-1582 (May 1975) "Gamma ray cascade decays from ψ(3684) to ψ(3095)".

HADRONIC MODELS

(104a)	H.P. Dürr	Phys. Rev. Lett. 34, 10 (1975) 422 "Is the ψ(3100) a $\Omega\Omega$ compound ? "
(104b)	H.P. Dürr	Phys. Rev. Lett. 34, 10 (1975) 616 "Are there narrow D-wave baryon-antibaryon resonances ? "
(104c)	C.T. Chen-Tsai T.Y. Lee	Taiwan Univ. (March 1975) " ψ (or J) particles as radially excited mesons."
(104d)	A.S. Goldhaber M. Goldhaber	Phys. Rev. Lett. 34, 1 (1975) 36 "Are the new particles paryon-antibaryon nuclei ? "
(104e)	O.D. Dalkarov V.B. Mandelzweig I.S. Shapiro	Nucl. Phys. B 21 (1970) 88 "On possible quasi-nuclear nature of heavy meson resonances."
(104f)	O.D. Dalkarov V.A. Khoze I.S. Shapiro	JETP Lett. 19, 8 (1974) 282 "Multiple generation of pions in colliding electron-positron beams."
(104g)	D. Weingarten S. Okubo	Phys. Rev. Lett. 34, 18 (1975) 1201 "Comment on a narrow resonance at 1932 MeV."

(104h)	D.M. Tow C.H. Tan K. Kang H.M. Fried	Phys. Rev. Lett. 34, 8 (1975) 499 "Conventional hadronic model of the new particle at 3.1 and 3.7 ."
(104i)	D.M. Tow	Brown University (April 1975) "New argument for the small widths of ψ's in the $\Omega\Omega$ model ."
(104j)	D.M. Tow	Brown University (May 1975) "Comments and critique of the $\Omega\bar{\Omega}$ model for the new resonances."
(104k)	C.A. Heusch	Lett. al Nuovo Cim. 12, 14 (1975) 552 "A remark on the baryon-antibaryon bound state model of the new narrow resonances."

CHARMONIUM

(105a)	J.H. Primack H.R. Quinn	Santa Cruz Summer School, 1973. "A practical introduction to gauge theories of weak interactions."
(105b)	J. Bernstein	Rev. Mod. Phys. 46, 1 (1974) 7 "Spontaneous symmetry breaking, gauge theories, the Higgs mechanism and all that."
(106a)	T. Appelquist H.D. Politzer	Phys. Rev. Lett. 34, 1 (1975) 43 "Heavy quarks and e^+e^- annihilation."
(106b)	T. Appelquist H.D. Politzer	FERMILAB-75/70-THY (Sept.1975) "Heavy quarks and their experimental consequences". (Proc. of Argonne Conf. July 1975)
(106c)	T.Appelquist	Caltech 68-499 (May 1975) "Narrow resonances and heavy quarks".
(106d)	J.M. Jauch F. Rohrlich	"The theory of photons and electrons", p. 282, Addison-Wesley Publ. Co. (1955).
(107a)	C.G. Callan R.L. Kingsley S.B. Treiman F. Wilczek A. Zee	Phys. Rev. Lett. 34, 1 (1975) 52 "Remarks on the new resonances at 3.1 GeV and 3.7 GeV ."
(107b)	A. De Rujula S.L. Glashow	Phys. Rev. Lett. 34, 1 (1975) 46 "Is bound charm found ? "
(107c)	J. Pasupathy G. Rajasekaran	Phys. Rev. Lett. 34, 19 (1975) 1250 "Comments on is bound charm found ? "

(107d)	A. de Rujula S.L. Glashow	Phys. Rev. Lett. 34, 19 (1975) 1252 "Response to the comment by and Rajasekaran."
(107e)	H.D. Politzer	Physics Reports 14 C, 4 (1974) "Asymptotic freedom : An approach to strong interactions."
(108)		See Refs . 73
(109a)	T. Appelquist A. De Rujula H.D. Politzer S.L. Glashow	Phys. Rev. Lett. 34, 6 (1975) 365 "Spectroscopy of the new mesons."
(109b)	F. Eichten K. Gottfried I. Kinoshita J. Kogut K.D. Lane T.M. Yan	Phys. Rev. Lett. 34, 6 (1975) 369 "Spectrum of charmed quark - anti quark bound states."
(109c)	B.J. Harrington S.Y. Park A. Yildiz	Phys. Rev. Lett. 34, 3 (1975) 168 "The spectra of heavy mesons in e^+e^- annihilation." E : Phys. Rev. Lett. 34, 5 (1975) 296
(109d)	T.P.Cheng P.B. James	Phys. Rev. Lett. 34, 14 (1975) 917 "How heavy are the quarks ? "
(109e)	H.J. Schnitzer	Harvard University (Jan. 1975) "Quark dynamics and bound charm."
(109f)	J.S. Kang H. J. Schnitzer	Brandeis Univ. (March 1975) "Dynamics of light and heavy bound quarks."
(109g)	B.J. Harrington S.Y. Park A. Yildiz	Phys. Rev. Lett. 34, 11 (1975) 706 "Orbital excitations in charmonium."
(109h)	P. Vinciarelli	Ref. TH. 1981 - CERN (Jan. 1975) "Radial recurrences in field theoretical models (and the spectrum of charmonium)."
(109i)	D.B.Lichtenberg J.G.Wills	Bloomington,Indiana (May 1975) "Spectrum of Strangeonium".
(109j)	A.P. Balachandran R. Ramachandran J. Schechter K.C. Wali H. Rupertsberger	Syracuse University (Febr. 1975) "Monopole strings and charmonium".

(109k) S. Iwao — Kanazawa University (April 1975)
"Classification of charmonium in terms of Elliot Model."

(109ℓ) H. Fritzsch, P. Minkowski — CALTECH (March 1975)
" ψ-resonances, Gluons and the Zweig rules."

(109m) M.B. Einhorn, C. Quigg — Fermilab - PUB - 75/0 - THY
"On the new narrow resonances."

(110) J. Borenstein, R. Shankar — Phys. Rev. Lett. 34, 10 (1975) 619
"Electromagnetic decay of the new heavy mesons at 3.1 and 3.7 GeV."

(111a) J. Kogut, L. Susskind — Phys. Rev. Lett. 34, 12 (1975) 767
"Electron-positron annihilation at and above the charm threshold for production of charmed hadrons."

(111b) R.Barbieri, R.Kögerler, Z.Kunszt, R.Gatto — Phys.Lett. 56B,5 (1975) 477
"Electron-positron annihilation above charm threshold".

ZWEIG'S SELECTION RULE

(112a) G. Zweig — Ref. TH. 401 and 412 - CERN (1964)

(112b) J.L. Rosner — Physics Reports 11 C (1974) 189
"The classification and decay of resonant particles."

(113a) Y. Iwasaki — Kyoto University (Dec. 1974)
"New meson at 3.1 GeV and quark duality diagram constraint."

(113b) See Refs. 73,109ℓ,65

(113c) F. Gilman — SLAC-PUB-1537 (Febr. 1975)
"Electron positron annihilation and the structure of hadrons."

(114) L. Clavelli — Maryland Univ. (Febr. 1975)
"Pomeron couplings and the decay of ψ and ψ' ."

(115) V.Blobel et al. — Bonn-Hamburg-Munich Collab. (1975)
"Test of Zweig selection rule in ϕ production by pp collisions".

RESONANCE MODELS, DUALITY

(116a) R.C. Brower, J.R. Primack — UC, Santa Cruz (Febr. 1975)
"Is the ψ a ring state ? "

(116b) R.C. Brower, J.R. Primack — UC, Santa Cruz (Dec. 1974)
"Is the ψ a ring ? "

(116c) K. Kikkawa, T. Kotani — Osaka Univ. (Febr. 1975) "A string model interpretation of ψ-particles."

(116d) A. Hosoya, T. Saito — Toyonaka Univ. (Dec. 1974) "Quantization for Regge slopes and ψ-particles."

(117) G. Cohen-Tannoudji, C. Gilain, G. Girardi, U. Maor, A. Morel — Saclay, Gif-sur-Yvette (March 1975) "A dynamical scheme for the decays of the ψ-particles."

(118) N.A. Törnqvist — Lett.al NC. 13 (1975) 341 "The relative hadronic widths of the psi mesons and $\psi \to D\bar{D} \to$ hadrons."

(119) M. Chaichian, J. Cleymans, R. Peschanski — Ref.Th. 1990 -CERN (March 1975) "Semi-local duality in e^+e^- annihilation into hadrons ? "

ZERO POLE SYSTEM

(120a) C.K. Chen — Glasgow Univ. (Dec. 1974) "Very narrow width resonance and CDD poles."

(120b) C.K. Chen — Phys.Rev.Lett. 35,1 (1975) 4 "New resonances in e^+e^- annihilation as the daughter of $\phi(1019)$."

(121) C.B. Chiu, R.J. Eden, M.B. Green — Phys. Rev. 185, 5 (1969) 1734 "Phase contours and bootstraps II."

(122) R.G. Newton — "Scattering theory of waves and particles." p. 607. Mc Graw-Hill Book Co.

NEW MEDIUM INTERACTIONS

(123a) C.W. Kim, A. Sato — Phys. Rev. Lett. 34, 6 (1975) 361 "Possible exsitence of new medium interactions and 3.1 and 3.7 GeV particles."

(123b) V.K. Cung, C.W. Kim, A. Sato — John Hopkins Univ. (Dec. 1974) "Phenomenology of the medium interactions."

(123c)	C.W. Kim A. Sato	John Hopkins Univ. (Dec. 1974) "Parity-violating medium interactions and their effects in nuclear physics and the 3.1 and 3.7 GeV particles."
(123d)	C.W. Kim A. Sato	John Hopkins Univ. (Jan. 1975) "Lepton coupling in the medium interaction and breakdown of muon-electron universality."
(124a)	J.C. Pati A. Salam	Phys. Rev. D 8, 4 (1973) 1240 "Unified lepton-hadron symmetry and a gauge theory of the basic interactions."
(124b)	J.C. Pati A. Salam	Phys. Rev. D 10, (1974) 275 "Lepton numbers as the fourth colour." E : Phys. Rev. D 10, (1974) 703
(124c)	J.C. Pati A. Salam	Phys. Rev. Lett. 34, 10 (1975) 613 "Are the new particles colour gluons ? "
(124d)	J.C. Pati	Maryland Univ. (Jan. 1975) "Particles, forces and the new mesons." (and references therein on this approach)
(125)	H. Fritzsch P.Minkowski	CALTECH (March 1975) " ψ-resonances, Gluons and the Zweig rule."
(126)	T.N. Pham B. Pire T.N. Truong	Phys. Rev. Lett. 34, 6 (1975) 347 "Remarks on the 3105 MeV resonance."
(127a)	N.T. Nieh T.T. Wu C.N. Yang	Phys. Rev. Lett. 34 , 1 (1975) 49 "Possible interactions of the J-particles."
(127b)	S.Y. Lo	Melbourne Univ., Australia (Dec. 1974) "Possible selection rules for the decay of J, ψ and θ ."
(128)	E. Everett	Phys. Rev. Lett. 34, 18 (1975) 1187 "Are the new e^+e^- resonances ordinary hadrons ? "
(129a)	R.E. Marshak R.N. Mohapatra	Phys. Rev. Lett. 34, 7 (1975) 426 "Strong W-pair model and the J-particles."
(129b)	R.E. Marshak R.N. Mohapatra	City College, N.Y. (May 1975) "Strong W-pair model and J particles."

VARIA

(130)	M.H. MacGregor	UC, Livermore (Jan. 1975) "The width of the $\psi(3105)$ as a scaling in $\alpha = e^2/hc$."
(131)	T. Das P.P. Divakaran L.K. Pandit V. Singh	Tata Inst. (Jan. 1975) "A test of models for the $\psi(3.1)$ using the branching ration into charged particles."

WEAK INTERACTION MODELS

(132a)	G. Altarelli N. Cabibbo R. Petronzio L. Maiani G. Parisi	Lett. al Nuovo Cim. 11, 14 (1974) 609 "Is the 3104 MeV vector meson the ϕ_c or the W_o ? "
(132b)	D. Weingarten	Rochester Univ. (Dec. 1974) "Tentative interpretation of the narrow e^+e^- annihilation resonance as a neutral weak vector boson."
(132c)	J.H. Reid	Calgary Univ. (Jan. 1975) "Could the narrow e^+e^- resonances be neutral weak vector bosons ? "
(132d)	S.C. Chhajlany	Liège Univ. (Jan. 1975) "On the possible interactions of neutral bosons with couplings to charged lepton pairs."
(132e)		See Refs. 65
(132f)	M. Glück E. Reya K. Schilcher	Phys. Lett. 56 B, 1 (1975) 73 "Is a low-mass Z^o in general SU(2) x U(1) gauge theories compatible with experiment ? "
(133a)	E.A. Paschos	Phys. Rev. Lett. 34, 6 (1975) 358 "Possible axial couplings of the new resonance."
(133b)	K.O. Michaelian	Phys. Rev. Lett. 34, 10 (1975) 611 "An analysis of the narrow resonance production in colliding beams."
(133c)	Min-Shih Chen Y.P. Yao	Phys. Rev. Lett. 34, 10 (1975) 628 "Comments on weak neutral vector mesons and charge asymmetry in $e^+e^- \to \mu^+\mu^-$."

(133d)	E. Lendvai K. Nagy G. Pocsik	Roland Eotvos Univ. (Nov. 1974) "Weak asymmetries in e^+e^- annihilation at high energies."
(133e)	J.D. Jackson	LRL, Berkeley (Dec. 1974) "On the effects of non-conservation of parity for a resonance in the channel $e^+e^- \rightarrow \mu^+\mu^-$."
(134)	S. Pakvasa G. Rajasekaran S.F. Tuan	Hawaii Univ. (Nov. 1974) "Theoretical implications of the resonance anomalies in e^+e^- system."
(135)	J.J.Sakurai	Phys. Rev. Lett. 34, 1 (1975) 56 "Intermediate boson in the fermion current model of neutral currents."
(136)	J. Schwinger	Phys. Rev. Lett. 34, 1 (1975) 37 "Interpretation of a narrow resonance in e^+e^- annihilation."
(137)	A. Soni	Stony Brook (Jan. 1975) "A gauge model for semileptonic interactions."
(138)	A.D. Dologov	Ref. TH. 1999 - CERN (March 1975) "On the possible interpretation of the new particles as gauge bosons."
(139a)	P.H.Frampton	Syracuse U. (April 1975) "Renormalizable model of leptons with new particles as neutral gauge bosons".
(139b)	P. H. Frampton	Syracuse Univ. (April 1975) "Model of leptons incorporating the new spin-one particles as gauge bosons."
(140)	W. Mc Kay H. Munczek	Phys. Rev. Lett. 34, 7 (1975) 432 "Higgs-particle interpretation of narrow e^+e^- resonances."

HEAVY PHOTON

(141a)	T.D. Lee G.C. Wick	Phys. Rev. D 2, (1970) 1033 "Finite theory of quantum electrodynamics."
(141b)	C. Marioni	Milano Univ. (Nov. 1974) "On finite gauge invariant quantum electrodynamics."
(142)	E. Etim A.F. Grillo G. Pancheri-Srivastava Y. Srivastava	Lett. al Nuovo Cim. 12, 2 (1975) 281 "Comments on the leptonic couplings of the $\psi(3.1)$ resonance."

Table I : Mass and decay widths of vector mesons

Particle	Mass (Gev)	Γ_{tot} (Mev)	Γ_e (kev)
ρ	0.770 ± 0.010	150 ± 0.010	6.5 ± 0.5
ω	0.7828± 0.0006	10 ± 0.4	0.76± 0.17
ϕ	1.0197± 0.0003	4.2 ± 0.2	1.34± 0.084
ψ(3.1)	3.095 ± 0.004	0.069 ± 0.015	4.8 ± 0.6
ψ'(3.7)	3.684 ± 0.005	0.225 ± 0.056	2.2 ± 0.3
ψ''(4.1)	4.15 ± 0.1	200 - 800	4 ± 1.2

Table III : Decay modes of ψ(3.7) (from Ref.10n,preliminary)

ψ(3.7) Decay Modes	B.R. (%)	Remarks
e^+e^-	0.97 ± 0.16	μ-e universality assumed
$\mu^+\mu^-$	0.97 ± 0.16	
ψ(3100) anything	57 ± 8	
ψ(3100) $\pi^+\pi^-$	32 ± 4	these decays are included in the fraction for ψ + anything
ψ(3100) η^o	4 ± 2	
ψ(3100) $\gamma\gamma$	<6.6*	via an intermediate state
$\rho^o\pi^o$	<0.1*	
$2\pi^+ 2\pi^-\pi^o$	<0.7*	
$p\bar{p}$	<0.03*	*preliminary analysis
hadrons	~ 20	

Table II : Decay modes of ψ(3.1)
(from Ref.10n , preliminary)

ψ(3.1) Decay Modes	B.R.(%)	Γ(kev)	Remarks
e^+e^-, $\mu^+\mu^-$	6.9 ± 0.9	4.8 ± 0.6	
$p\bar{p}$	0.21 ± 0.04		
$\Lambda\bar{\Lambda}$	0.16 ± 0.08		
$p\bar{p}\ \pi^0$ }			
$\bar{n}p\ \pi^-$ }	0.37 ± 0.19		
$\bar{p}n\ \pi^+$ }			
$\eta\gamma$	0.15 →1.8		
$2\pi^+2\pi^-$	0.4 ± 0.1		via γ
$2\pi^+2\pi^-K^+K^-$	0.3 ± 0.1		
$2\pi^+2\pi^-\pi^0$	4.0 ± 1.0		{ 20% $\omega\pi^+\pi^-$ 30% $\rho\pi\pi\pi$
$\pi^+\pi^-K^+K^-$	0.4 ± 0.2		→ not included K*(892) , K*(1420)
$\pi^+\pi^-p\ \bar{p}$	seen		
$\pi^+\pi^-\pi^0 \sim \rho\pi$	∿ 1.5		
$3\pi^+3\pi^-$	0.4 ± 0.2		via γ
$3\pi^+3\pi^-\pi^0$	2.9 ± 0.7		
$4\pi^+4\pi^-\pi^0$	0.9 ± 0.3		
$\omega\pi\pi$	0.9 ± 0.3		
$\rho\pi$	1.3 ± 0.3		
K_SK_L	< 0.02		
$K^oK^{o*}(892)$	0.24 ± 0.05		
$K^{\pm}K^{\mp*}(892)$	0.31 ± 0.07		
$K^oK^{o*}(1420)$	< 0.19		
$K^{\pm}K^{\mp*}(1420)$	< 0.19		
$K^{*o}(892)\bar{K}^{*o}(892)$	< 0.06		
$K^{*o}(1420)\bar{K}^{*o}(1420)$	< 0.18		
$K^{*o}(892)\ K^{*o}(1420)$	0.37 ± 0.10		
$\pi^+\pi^-$	not seen		G-violation
K^+K^-	not seen		
$\gamma\gamma$	not seen		C-violation
$\pi^o\gamma$	small		

Table IV : Photoproduction of vector mesons: cross section and coupling constant (from Ref.15c)

Particle	Photon Energy (Gev)	$\sigma(\gamma N\to VN)$ (nb)	b $(Gev/c)^{-2}$	$g_V^2/4\Pi$	$\sigma(VN)$ (mb)	Refs.
ρ^0	9.3	13500±500	6.5±0.2	2.3±0.3	23±3	(see 15c)
ω	9.3	1800± 70	6.6±1.1	18.4±1.8	24±3	(see 15c)
ϕ	9.3	550± 70	4.6±0.7	12.2±1.0	9±1	(see 15c)
$\psi(3.1)$	11.1	0.65	5	2.8	0.82	(Cornell) (15a,b)
$\psi(3.1)$	18.2	$3.7^{+2.2}_{-1.5}$	4	13±4	1.0±0.3	(SLAC) (15c,d,e)
$\psi(3.1)$	<200				∿1	(FNAL) (15f)

Table V : Characteristics of modified forms of the Fermi model

Models (f_i =)	multi- plicity $\langle n\rangle$	mean momentum $\langle p\rangle$	n-particle cross section σ_n
$2E_i$	$(q^2)^{3/8}$	$(q^2)^{1/8}$	
1	$(q^2)^{1/3}$	$(q^2)^{1/6}$	$b_n \cdot (q^2)^{n-2}$
$1/E_i^2$	$\ln q^2$	$\sqrt{q^2}/\ln q^2$	
$e^{-\alpha \cdot p_{t\,i}^2}$	$\ln q^2$	$\sqrt{q^2}/\ln q^2$	
$e^{-\alpha \cdot \lvert\vec{p}_i\rvert}$ $\sim e^{-\alpha \cdot E_i}$	$\sqrt{q^2}$	const.	$e_n \cdot (q^2)^{n-3} \cdot e^{-a\sqrt{q^2}}$
$(1/E_i)^r$ r>2	const.	$\sqrt{q^2}$	

Table VI : Characteristics of parton models and new interaction models

PARTON - MODELS	$<n>$	$<p>$	$d\sigma/d\cos\Theta$	$q^2 \cdot \frac{d\sigma}{dx}$	σ_h	R_∞	Parameters
Spin-0	$\ln q^2$	$(q^2)^{1/2}$	$1-\cos^2\Theta$ (jets)	$\sum_i Q_i \cdot D_i(x)$	$\frac{\pi}{3}\frac{\alpha^2}{q^2}(\Sigma Q_i^2)_0$	$\frac{1}{4}(\Sigma Q_i^2)_0$	$Q_i, D_i(x)$
Spin-½	$\ln q^2$	$(q^2)^{1/2}$	$1+\cos^2\Theta$ (jets)	$\sum_i Q_i \cdot D_i(x)$	$\frac{4\pi}{3}\frac{\alpha^2}{q^2}(\Sigma Q_i^2)_{1/2}$	$(\Sigma Q_i^2)_{1/2}$	$Q_i, D_i(x)$
Spin-1	$\ln q^2$	$(q^2)^{1/2}$	$1+\cos^2\Theta$ (jets)	$\frac{1}{\sigma}\cdot\frac{d\sigma}{dx} \to$ fct.(x)	$\frac{\pi\alpha^2}{3} M_V^2$	$(\frac{q^2}{4M_V^2})$	$Q_i, D_i(x)$
Yang-Mills	$\ln q^2$	$(q^2)^{1/2}$	$1+\cos^2\Theta$ (jets)	$\sum_i Q_i \cdot D_i(x)$	$\frac{3\pi\alpha^2}{4} \cdot \frac{1}{q^2}$	$\frac{9}{16}$	$Q_i, D_i(x)$
structured	$\ln q^2$	$(q^2)^{1/2}$	$\{(1+\cos^2\Theta) + \mu_Q^2 \cdot q^2 (1-\cos^2\Theta)\}$	$\lvert F\rvert^2\{A(x)+\mu_Q^2 \cdot q^2 \cdot B(x)\}$ $\frac{1}{\sigma}\cdot\frac{d\sigma}{dx} \to$ fct.(x)	$\approx$ const.	$\lvert F\rvert^2 \cdot (1+q^2\mu_Q^2)$	$Q_i, D_i(x)$ $\mu_Q, F(q^2)$
NEW INTERACTION -MODELS	$<n>$	$<p>$	$d\sigma/d\cos\Theta$	$q^2 \cdot \frac{d\sigma}{dx}$	σ_h	R_∞	Parameters
Hadronic Leptons	$\ln q^2$	$(q^2)^{1/2}$	q^2 $\Theta=0,\pi$ const. $\Theta=\pi/2$	$\frac{d\sigma}{dx}$ (see text)	$2\pi G_F$	$\frac{3}{2}\frac{G_F}{\alpha^2} q^2$	$E \cdot \frac{d^3\sigma}{dp^3}$ (Ansatz)
Lepto- Hadron X	$\ln q^2$	$(q^2)^{1/2}$	$\to\{4+(1+\cos\Theta)^2\}$	$\frac{1}{32\pi}\sum_i D_i(x) \cdot f_i(q^2)$ $f_i = (aq^4+bq^2+c+d\frac{1}{q^2})_i$	$\frac{4\pi}{3}\frac{\alpha^2}{q^2} R_\infty$	$(\Sigma Q_i^2)_{1/2} \cdot \{1+ 2\delta_1 \cdot q^2 + \delta_2 \cdot q^4\}$	g_i = coupling const. $D_i(x)$

Table VII : Characteristics of vector meson dominance models

VECTOR-MESON MODELS	$<n>$	$<p>$	$\frac{d\sigma}{d\cos\theta}$	$q^2 \cdot \frac{d\sigma}{dx}$	σ_h	R_∞	Parameters
VMD	$\ln q^2$ (supposed)	$\sqrt{q^2}$	const. (supposed)	$\sim 1/q^2$	$\frac{12\pi}{q^4} \Gamma_\rho \cdot \Gamma_{\rho\to e^+e^-}$	$\frac{9}{\alpha^2} \Gamma \cdot \Gamma_\rho \, _{\rho\to e^+e^-} \cdot \frac{1}{q^2}$	$\Gamma_\rho, \Gamma_{\rho\to e^+e}$
EVMD (scaling)	$\ln q^2$ (supposed)	$\sqrt{q^2}$	const. (supposed)	$q^2 \frac{d\sigma}{dx} = \text{fct.}(x)$	$\frac{4\pi}{3} \frac{\alpha^2}{q^2} \cdot R_\infty$	$R_\infty = 2 - 3$	$m_n^2 = m_\rho^2 (1+a\cdot n)$ Γ_n, f_n
EVMD (non-scaling)	$\ln q^2$ (supposed)	$\sqrt{q^2}$	const. (supposed)	$\frac{1}{\sigma} \frac{d\sigma}{dx} \to \text{fct.}(x)$	16 nbarn	const. $\frac{1}{q^2}$	$m_n^2 = m_\rho^2 (1+a\cdot n)$ Γ_n, f_n
EVMD (truncated)	$\ln q^2$ (supposed)	$\sqrt{q^2}$	const. (supposed)	$q^2 \cdot \frac{d\sigma}{dx} \to \text{fct.}(x)$	$\frac{3\pi N}{4m_\rho^2 q^2} \cdot \Gamma_\rho \cdot \Gamma_{\rho\to e^+e^-}$	$\frac{9N}{16\alpha^2 m_\rho^2} \cdot \Gamma_\rho \cdot \Gamma_{\rho\to e^+e^-}$	$f_n(q^2), \Gamma_n(q^2)$ N
CUT - MODEL	$\ln q^2$ (supposed)	$\sqrt{q^2}$	$1 \pm \cos^2\theta$ (jets)	$\frac{1}{\sigma} \frac{d\sigma}{dx} = \text{fct.}(x)$	const.	$\sim q^2$	$m_n^2 = m_\rho^2 (1+a\cdot n)$ $F_n \equiv F(q^2, m_n^2)$

Table VIII : Characteristics of statistical models

STATISTICAL MODELS	$\langle n\rangle$	$\langle p\rangle$	$\frac{d\sigma}{d\cos\theta}$	$\frac{1}{\sigma_{tot}}\cdot\frac{d\sigma}{dx}$	σ_n/σ_{tot}	Parameters
Fermi	$(q^2)^{1/3}$	$(q^2)^{1/6}$	const.	$q^2\cdot e^{h(x)\cdot q^{2/3}}$	$a_n(q^2)^{2n-8/3}\cdot e^{-\alpha\sqrt{q^2}}$	$K \to V_o=$const.
Landau - 1	$(q^2)^{3/8}$	$(q^2)^{1/8}$	const.	$(q^2)^{3/8}$	-	$V_o=$const.
Landau - 2	const.	$(q^2)^{1/2}$	const.	?	-	$V_o=(q^2)^{-2/3}$
Heisenberg	?	?	const.	?	-	V_o
Thermodyn.	$(q^2)^{1/2}$	const.	const.	$f(\beta)\cdot(q^2)^{\frac{3}{2}}\cdot e^{-\beta\frac{\sqrt{q^2}}{2}\cdot x}$	$b_n\cdot(q^2)^{n-\frac{5}{4}}\cdot e^{-\beta\sqrt{q^2}}$	$\beta=kT_o$
Uncorr.Jet	$\ln q^2$	$(q^2)^{1/2}$	const.+jets	$\sim 6\cdot\frac{(1-x)^2}{x}$	$c_n(q^2)^{-\bar{K}}\cdot(\ln\frac{q^2}{4m^2})^{n-1}$	K,λ
Cascade	$(q^2)^{-\alpha(O)}$	$(q^2)^{1+\alpha(o)}$	const.	$\sim e^{-\frac{(y-\bar{y})^2}{2y_o}}$	$\frac{1}{\langle n\rangle}\Phi(\frac{n}{\langle n\rangle})$	Probability:P_n Fireball→Particles

Table IXa : Quantum number assignment to quarks using: $Q = I_3 + \frac{1}{2}(S+B+C)$

Quarks	p	n	λ	c	$\bar{p}$	$\bar{n}$	$\bar{\lambda}$	$\bar{c}$
B	$\frac{1}{3}$	$\frac{1}{3}$	$\frac{1}{3}$	$\frac{1}{3}$	$-\frac{1}{3}$	$-\frac{1}{3}$	$-\frac{1}{3}$	$-\frac{1}{3}$
Q	$\frac{2}{3}$	$-\frac{1}{3}$	$-\frac{1}{3}$	$\frac{2}{3}$	$-\frac{2}{3}$	$\frac{1}{3}$	$\frac{1}{3}$	$-\frac{2}{3}$
Y	$\frac{1}{3}$	$\frac{1}{3}$	$\frac{2}{3}$	$\frac{1}{3}$	$-\frac{1}{3}$	$-\frac{1}{3}$	$+\frac{2}{3}$	$-\frac{1}{3}$
S	0	0	-1	0	0	0	+1	0
I	$\frac{1}{2}$	$\frac{1}{2}$	0	0	$\frac{1}{2}$	$\frac{1}{2}$	0	0
I_3	$\frac{1}{2}$	$-\frac{1}{2}$	0	0	$-\frac{1}{2}$	$\frac{1}{2}$	0	0
C	0	0	0	1	0	0	0	-1

Table IXb,c : Quantum number assignment to quarks using

$Q = I_3 + \frac{1}{2}Y + \frac{2}{3}C$ (case b)) or $Q = I_3 + \frac{1}{2}(S+B) + C$ (case c))

Quarks	p	n	λ	$c^{(b)}$	$c^{(c)}$
B	$\frac{1}{3}$	$\frac{1}{3}$	$\frac{1}{3}$	$\frac{1}{3}$	$\frac{1}{3}$
Q	$\frac{2}{3}$	$-\frac{1}{3}$	$-\frac{1}{3}$	$\frac{2}{3}$	$\frac{2}{3}$
Y	$\frac{1}{3}$	$\frac{1}{3}$	$\frac{2}{3}$	0	$-\frac{2}{3}$
S	0	0	-1	$-\frac{1}{3}$	-1
I	$\frac{1}{2}$	$\frac{1}{2}$	0	0	0
I_3	$\frac{1}{2}$	$-\frac{1}{2}$	0	0	0
C	0	0	0	1	1

Table X : Expected spectrum of hidden charmed mesons

Meson	Quark content	Spin-parity	Mass (MeV)
η_c	$c\bar{c}$	0^-	3000?
ϕ_c	$c\bar{c}$	1^-	3100
ε_c	$c\bar{c}$	0^+	3200??
A_c	$c\bar{c}$	1^+	3300?
f_c	$c\bar{c}$	2^+	3500?
g_c	$c\bar{c}$	3^-	3700?
ϕ_c	$c\bar{c}$	1^-	3700

} Hidden charmed mesons

Table XI : Expected spectrum of charmed mesons

Meson	Quark content	Spin-parity	Mass (MeV)
D	$c\bar{p}, c\bar{n}$	0^-	$m_D = 0(2000)$
F	$c\bar{\lambda}$	0^-	$m_D + 60?$
D*	$c\bar{p}, c\bar{n}$	1^-	$m_D + 120?$
F*	$c\bar{\lambda}$	1^-	$m_D + 180?$
D_s	$c\bar{p}, c\bar{n}$	0^+	$m^*_D = m_D + 150??$
F_s	$c\bar{\lambda}$	0^+	$m^*_F = M_D + 200??$

} Charmed mesons

Table XII : Influence of η_c-mass on radiative decay width of ϕ_c

$\Gamma(\phi_c \to \eta_c \gamma)$ (kev)	m_{η_c} (Gev)
100	2.7
6	2.95
0.3	3.05

Table XIII : Radiative decay widths of ρ,ω,ϕ

Decay	Experimental width(Kev)
$\omega\to\gamma\pi$	890 ± 90
$\rho\to\gamma\pi$	<730
$\phi\to\gamma\pi$	<14
$\rho\to\gamma\eta$	<160
$\omega\to\gamma\eta$	<49
$\phi\to\gamma\eta$	126 ± 46
$X^o\to\gamma\rho$	$0.26\cdot\Gamma(X^o\to \text{all})$
$X^o\to\gamma\omega$	
$\phi\to\gamma X^o$	

Table XIV : Two photon decay widths of π^o and η,η' -mesons

$$\Gamma_{\pi^o\to\gamma\gamma} = 7.8 \pm 0.9 \text{ ev}$$

$$\Gamma_{\eta\to\gamma\gamma} = 374 \pm 60 \text{ ev}$$

$$\Gamma_{\eta'\to\gamma\gamma} \leqq 19 \pm 3 \text{ Kev}$$

Table XV : Proposed Color explanations of the ψ-particles

Symmetry		ψ	ψ'	ψ''	Ref.
$SU_3 x SU_3^c$		$(1,\gamma)$	radial+orbital excitations		(82)
$SU_3 x SU_3^c$		$(1,\rho^o)$	$(1,\omega_8)$	do.	(83m)
$SU_3 x SU_3^c$		(ω,γ)	(ϕ,γ)	(ω,γ)-recurrence (or threshold)	(85)
$SU_3 x SU_3^c$		(ω,ρ^o)	(ω,ω_8)	(ϕ,ρ^o) (ϕ,ω_8) anticipated	(83 o,p; 84)
$SU_3 x SU_3^c$		(ω,γ)	(ω,γ)-recurrence	(ϕ,γ)	(83q)
$SU_3 x SU_3^c$		$A\bar{A}$	$\sqrt{\frac{1}{2}}\{B_1\bar{B}_1 + B_2\bar{B}_2\}$	$\sqrt{\frac{1}{3}}\{c_4\bar{c}_4 + c_1\bar{c}_1$ $c_{-2}\bar{c}_{-2}\}$	(82a)
$SU_3 x SU_3^c$		$SU_3 X SU_2(I^c) x SU_1(Y^c)$			(83a,b)
$SU_4 x SU_3$					(87)

Table XVI : Quantum number assignment to quarks in the $SU_3 x SU_1(z)$ scheme (Ref.90)

Quarks	p	n	λ	χ
B	$\frac{1}{3}$	$\frac{1}{3}$	$\frac{1}{3}$	$\frac{1}{3}$
Q	$\frac{2}{3}$	$-\frac{1}{3}$	$-\frac{1}{3}$	$-\frac{1}{3}$
Y	$\frac{1}{3}$	$\frac{1}{3}$	$-\frac{2}{3}$	$-\frac{2}{3}$
S	0	0	-1	-1
I	$\frac{1}{2}$	$\frac{1}{2}$	0	0
I_3	$\frac{1}{2}$	$-\frac{1}{2}$	0	0
Z	0	0	0	1

Table XVII : Quantum number assignment to leptons in the Konopinski-Mahmoud scheme (Ref.97)

Particle	e^-	ν_e	μ^-	$\bar{\nu}_\mu$	e^+	$\bar{\nu}_e$	μ^+	ν_μ
L_e	1	1	0	0	-1	-1	0	0
L_μ	0	0	1	-1	0	0	-1	1
L	1	1	-1	1	-1	-1	+1	-1
(λ)		-1		+1		+1		-1

Table XVIII : Introduction of a new lepton quark ℓ - quantum number assignment (Ref.97)

Particle	e^-	ν	μ^-	ℓ
I_3	$-\frac{1}{2}$	$+\frac{1}{2}$	0	0
L	1	1	-1	-1
12 C	3	3	3	-9
Q	-1	0	-1	+2

Table XIX : Quantum number assignment to leptons similar to hadrons (Ref. 99)

Particle	e^-	ν_e	μ^-	$\bar{\nu}_\mu$	e^+	$\bar{\nu}_e$	μ^+	ν_μ
I	$\frac{1}{2}$	$\frac{1}{2}$	$\frac{1}{2}$	$\frac{1}{2}$	$\frac{1}{2}$	$\frac{1}{2}$	$\frac{1}{2}$	$\frac{1}{2}$
I_3	$-\frac{1}{2}$	$+\frac{1}{2}$	$-\frac{1}{2}$	$+\frac{1}{2}$	$+\frac{1}{2}$	$-\frac{1}{2}$	$+\frac{1}{2}$	$-\frac{1}{2}$
ℓ	+1	+1	-1	-1	-1	-1	+1	+1
S	-2	-2	0	0	+2	+2	0	0
Q	-1	0	-1	0	+1	0	+1	0

Table XX : Quantum number assignment to gauge particles in a phenomenological theory for weak and strong interactions (Ref. 129)

Particle	W^+	W^0	$W^{0'}$
Q	1	0	0
I	$\frac{1}{2}$	$\frac{1}{2}$	0
I_3	$\frac{1}{2}$	$-\frac{1}{2}$	0
t	1	1	1
$\bar{Y}$	1	1	0
Y	$\frac{1}{3}$	$\frac{1}{3}$	$-\frac{2}{3}$

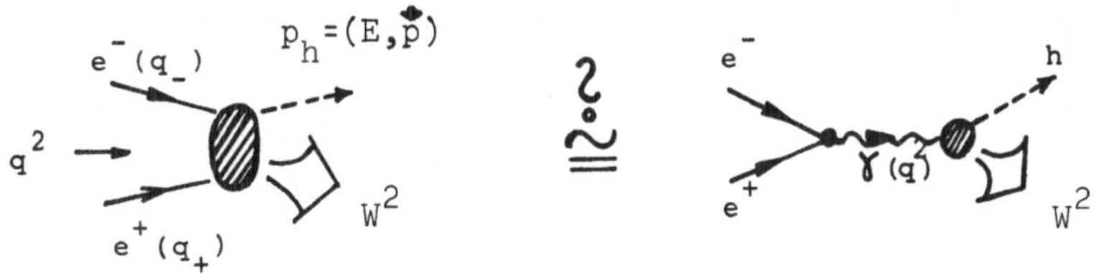

Figure 1 : Inclusive hadron production in e^+e^- annihilation via single-photon exchange

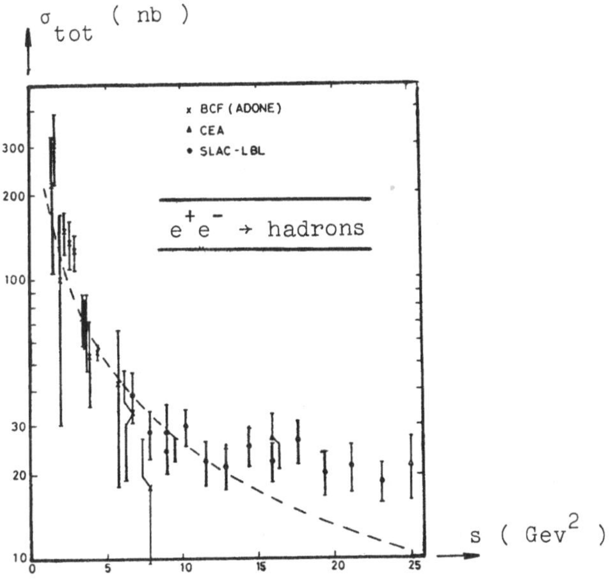

Figure 2 : Total cross-section σ_h of $e^+e^-\to$ hadrons in the range 3 Gev $\leq E_{CM} \leq$ 5 Gev (from Ref.1)

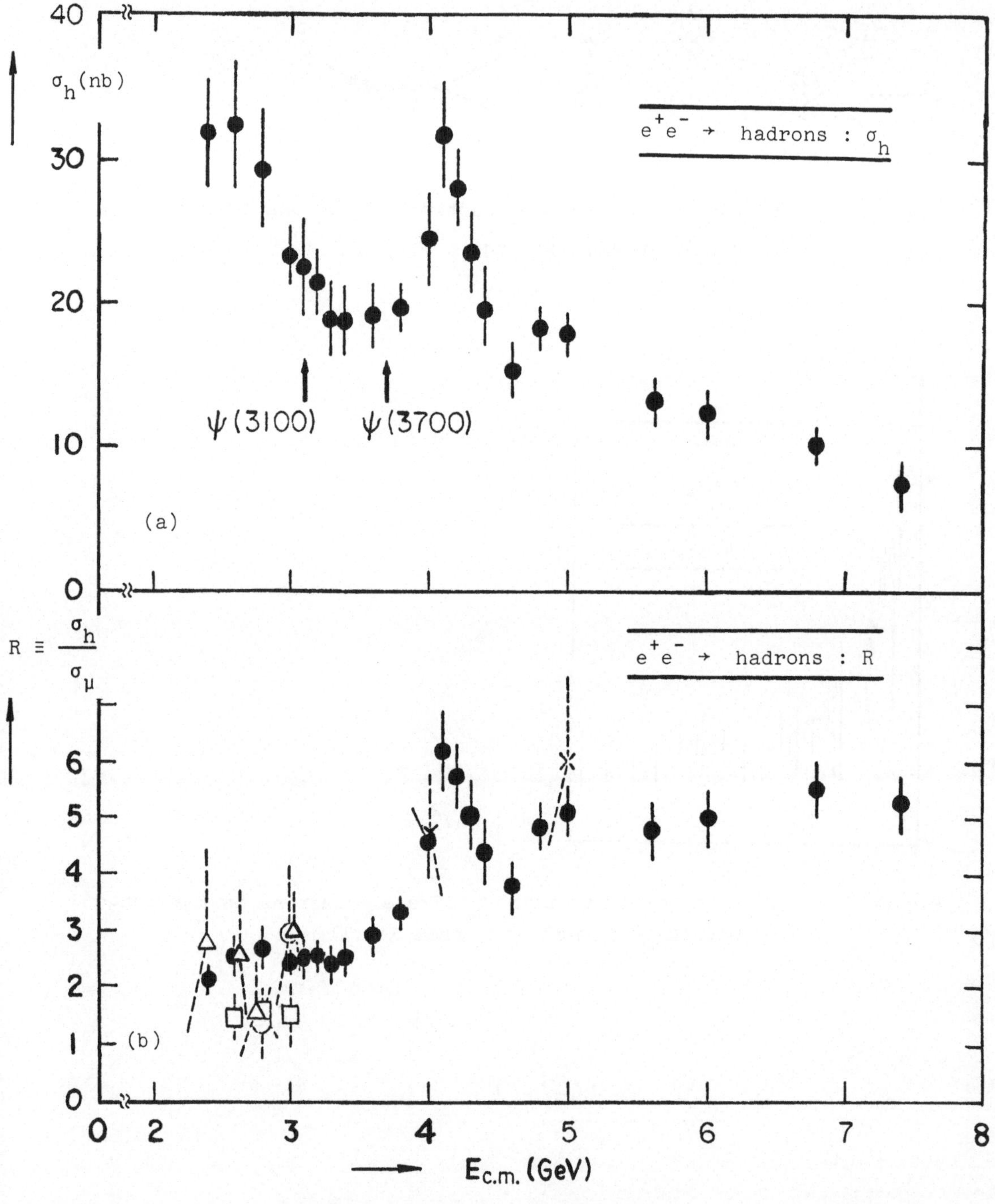

Figure 3 : Total cross section σ_h and ratio R of $e^+e^- \to$ hadrons in the range 3 Gev $\leq E_{CM} \leq$ 7 Gev (from Ref.10 o)

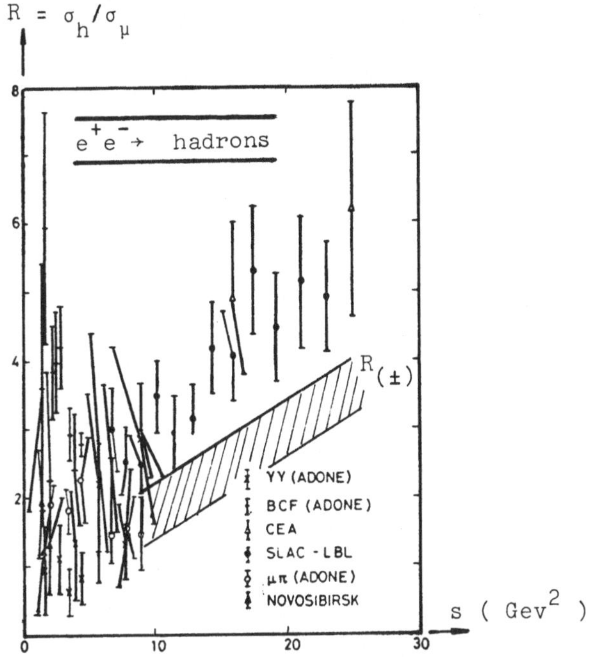

Figure 4 : Ratio of total cross-sections for hadron and lepton production in e^+e^- annihilation: $R = \sigma_h/\sigma_\mu$ (from Ref.1)

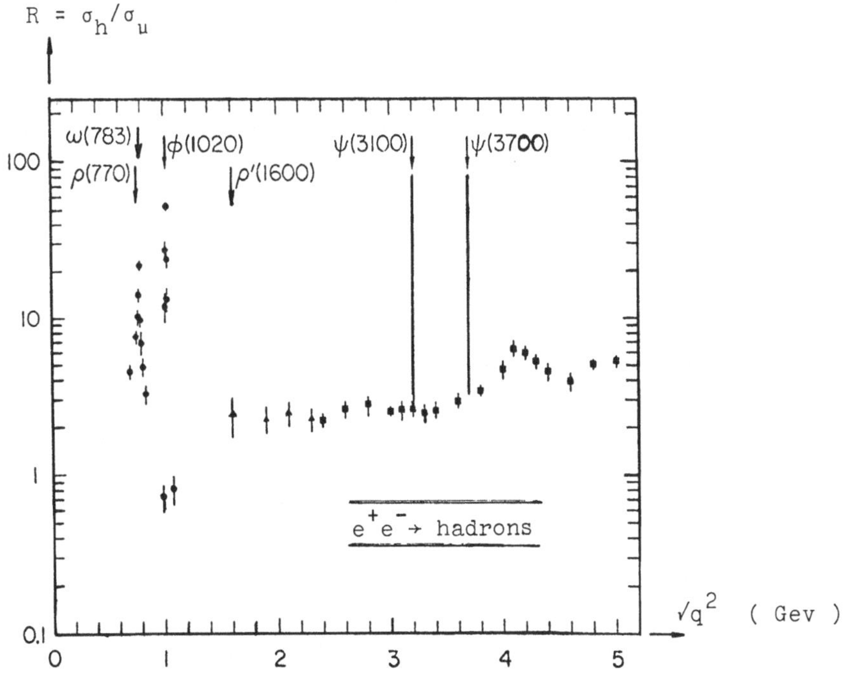

Figure 5 : Ratio of total cross-sections for hadron and lepton production in e^+e^- annihilation: $R = \sigma_h/\sigma_\mu$ on a logarithmic scale (from Ref.5)

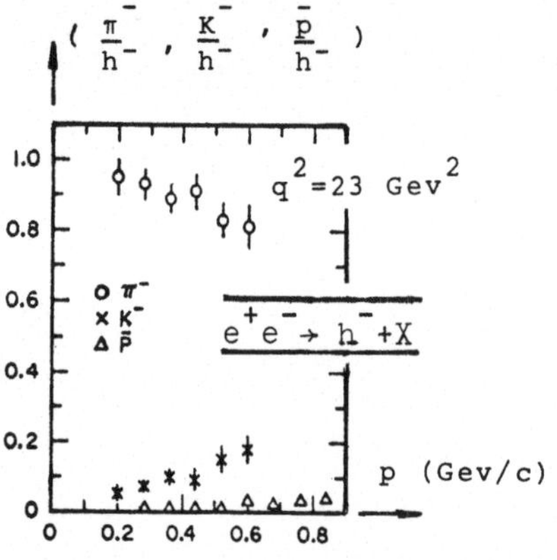

Figure 6 : Ratios $\{\pi^-/_h(-),\ K^-/_h(-),\ \bar{p}/_h(-)\}$ as functions of $E_{CM} \equiv \sqrt{q^2}$. $h^{(-)} \equiv$ {all negatively charged particles}(from Ref.1)

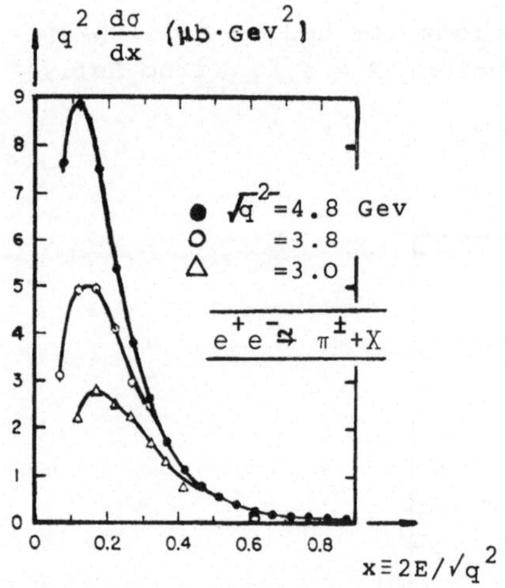

Figure 7 : $e^+e^- \to h^{(\pm)} + X$: Inclusive distribution $q^2 \cdot \frac{d\sigma}{dx}$ vs. $x \equiv \frac{2E}{\sqrt{q^2}}$ = 3.0, 3.8, 4.8 Gev (from Ref.1)

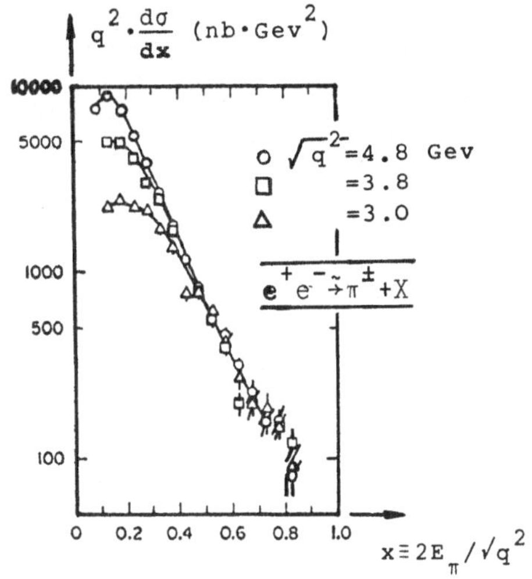

Figure 8 : $e^+e^- \to h^{(\pm)} + X$: Inclusive distribution $q^2\, d\sigma/dx$ vs. x on a logarithmic scale at $\sqrt{q^2}$ = 3.0,3.8,4.8 Gev (from Ref.1)

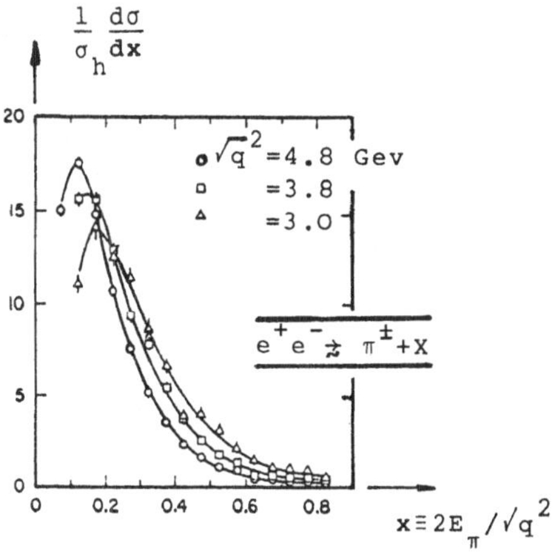

Figure 9 : $e^+e^- \to h^{(\pm)} + X$: Inclusive distribution $\frac{1}{\sigma_h} \cdot \frac{d\sigma}{dx}$ for q^2 = 3.0, 3.8, 4.8 Gev (from Ref.1)

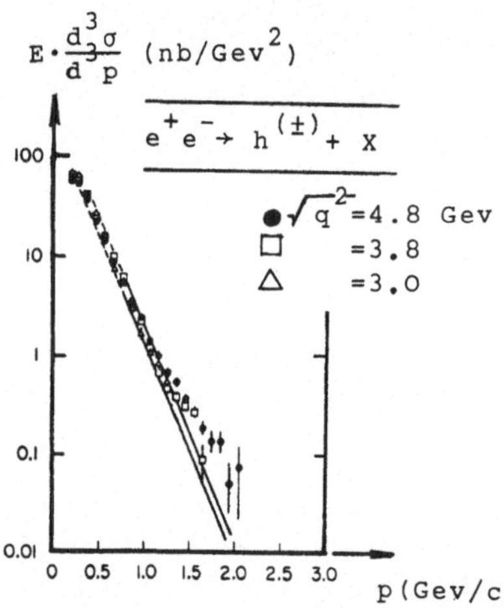

Figure 10 : Inclusive cross section $\left(E \frac{d^3\sigma}{dp^3}\right)$ vs. p at CM-energies $\sqrt{q^2}$ = 3.0, 3.8, 4.8 Gev (from Ref.1)

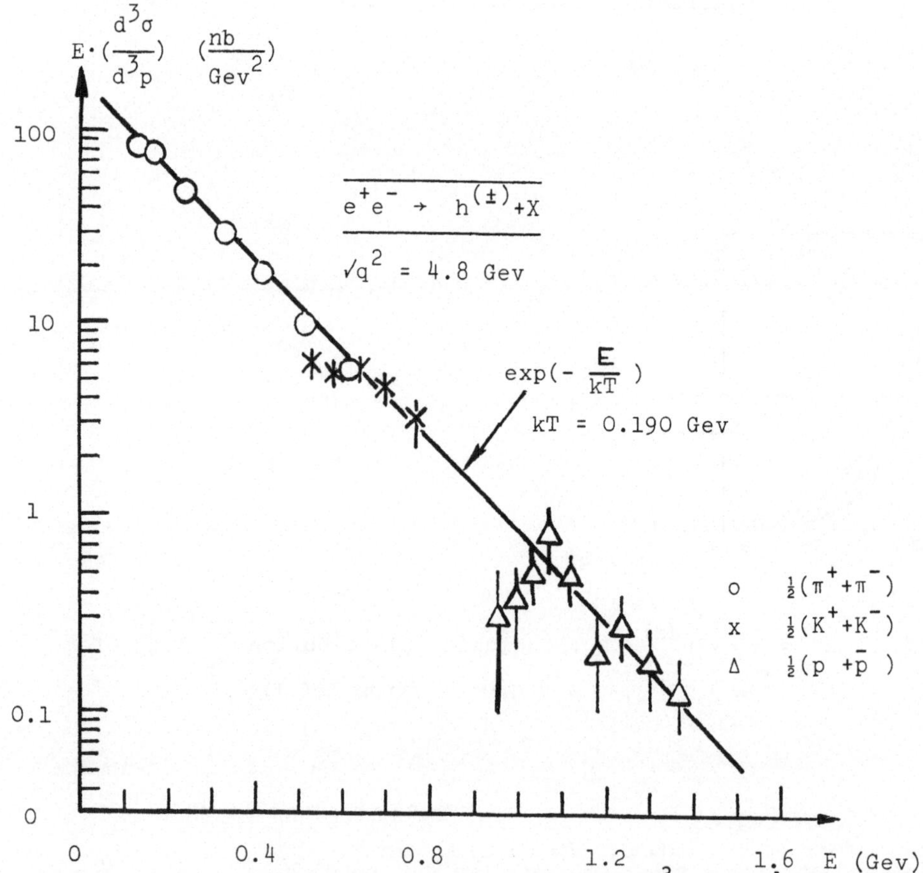

Figure 11 : Inclusive cross section $E \cdot \frac{d^3\sigma}{dp^3}$ vs. E of (π^-, K^-, p^-) as inclusive particles at $q^2 = 23\ \mathrm{Gev}^2$ (from Ref.10-o)

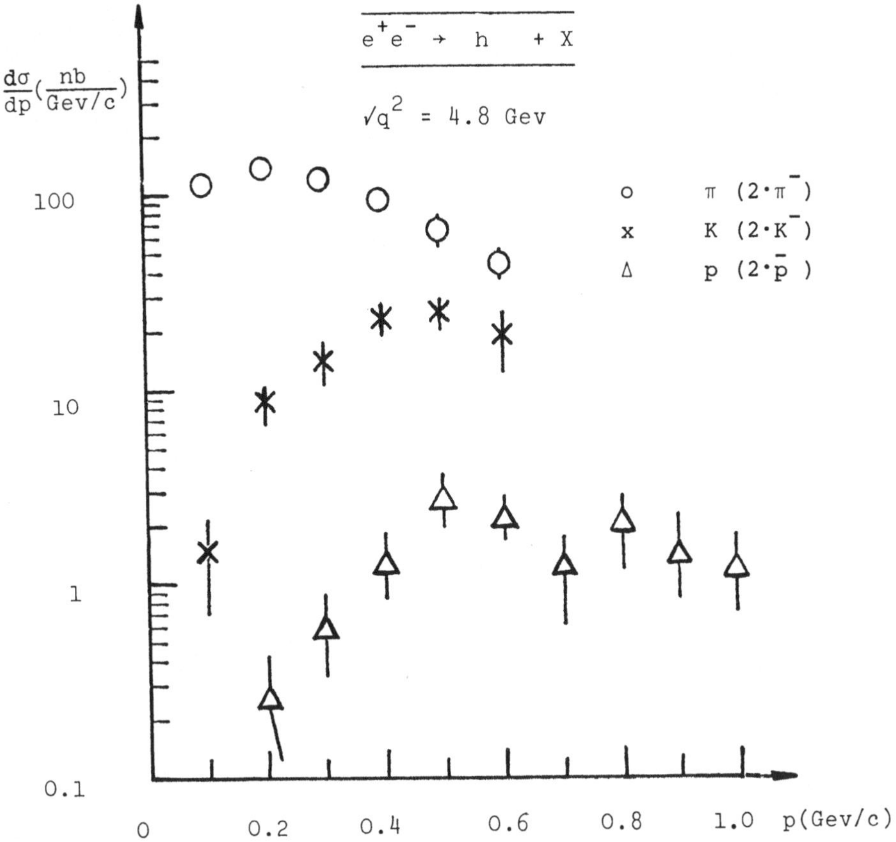

Figure 12 : Inclusive momentum distribution of charged particles π^-, K^-, $\bar{p}$ (from Ref. 10p)

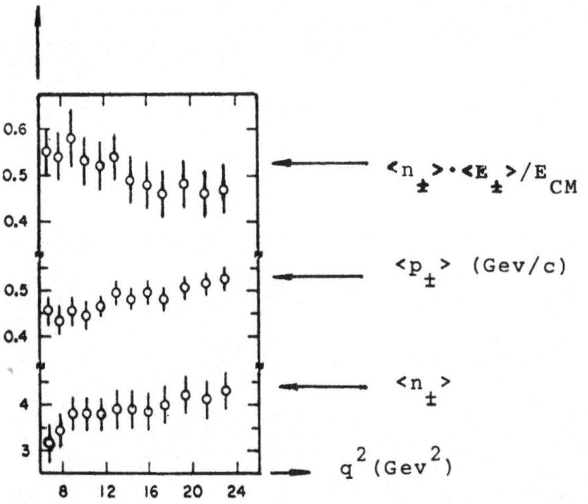

Figure 13a : - Mean charged particle multiplicity : $<n_{\pm}>$ vs. q^2 .
- Mean charged particle momentum : $<p_{\pm}>$ vs. q^2 .
- Ratio of charged $<E_{\pm}> \equiv <n_{\pm}> \cdot <p_{\pm}>$ to total initial energy $E_{CM} \equiv \sqrt{q^2}$: $<E_{\pm}>/E_{CM}$ vs. q^2 (from Ref.1a)

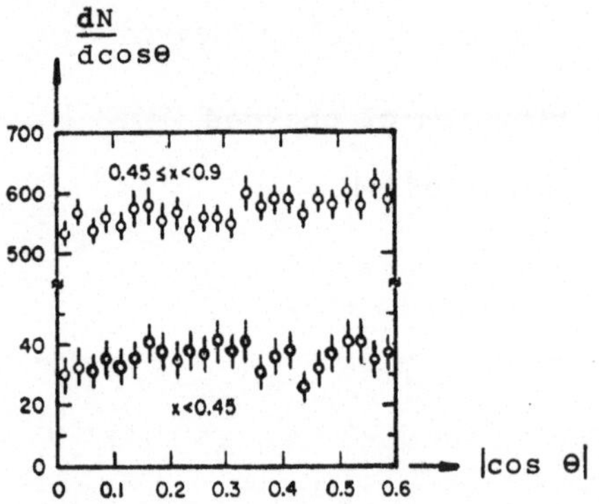

Figure 14 : Angular distribution $d\sigma/d\cos\Theta$ vs. $\cos\theta$ of the inclusive process $e^+e^- \to h^{(\pm)} + X$ (from Ref.1a)

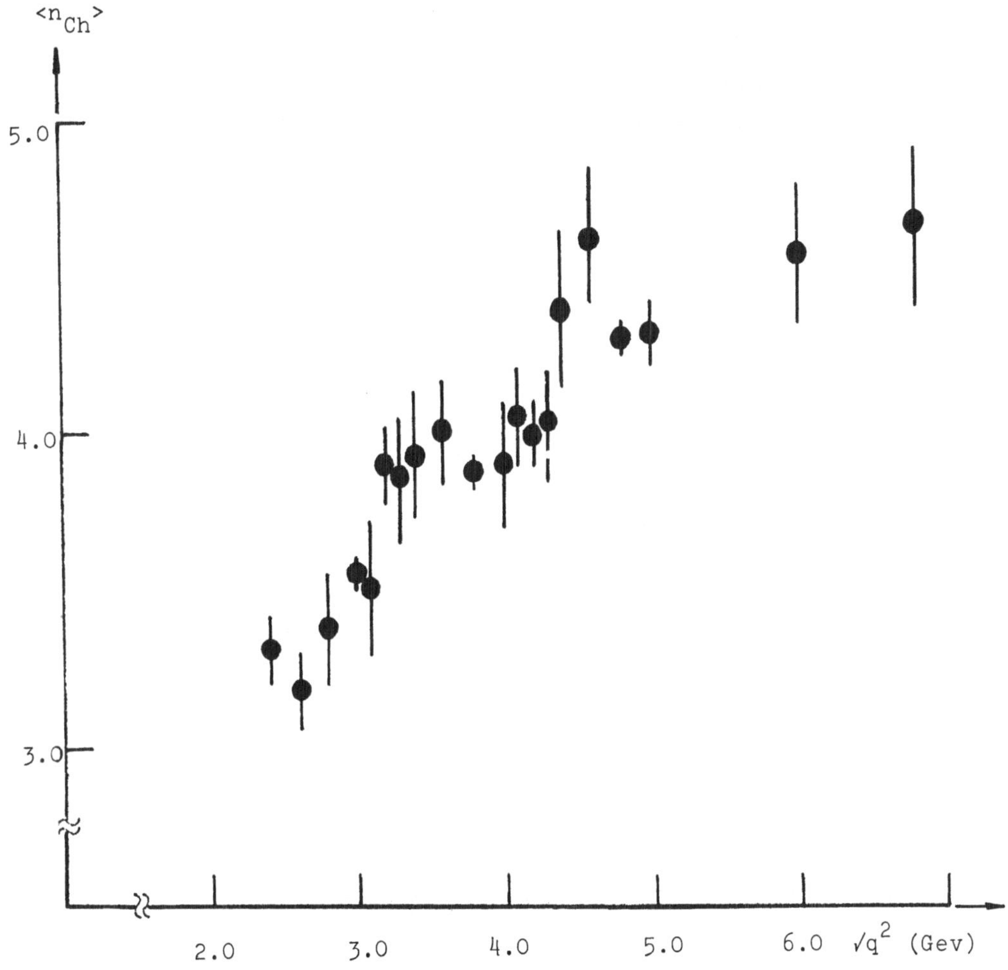

Figure 13b : Mean charge particle multiplicity : $<n_{\pm}>$ vs. $\sqrt{q^2}$ (more recent data) (from Ref.10-o)

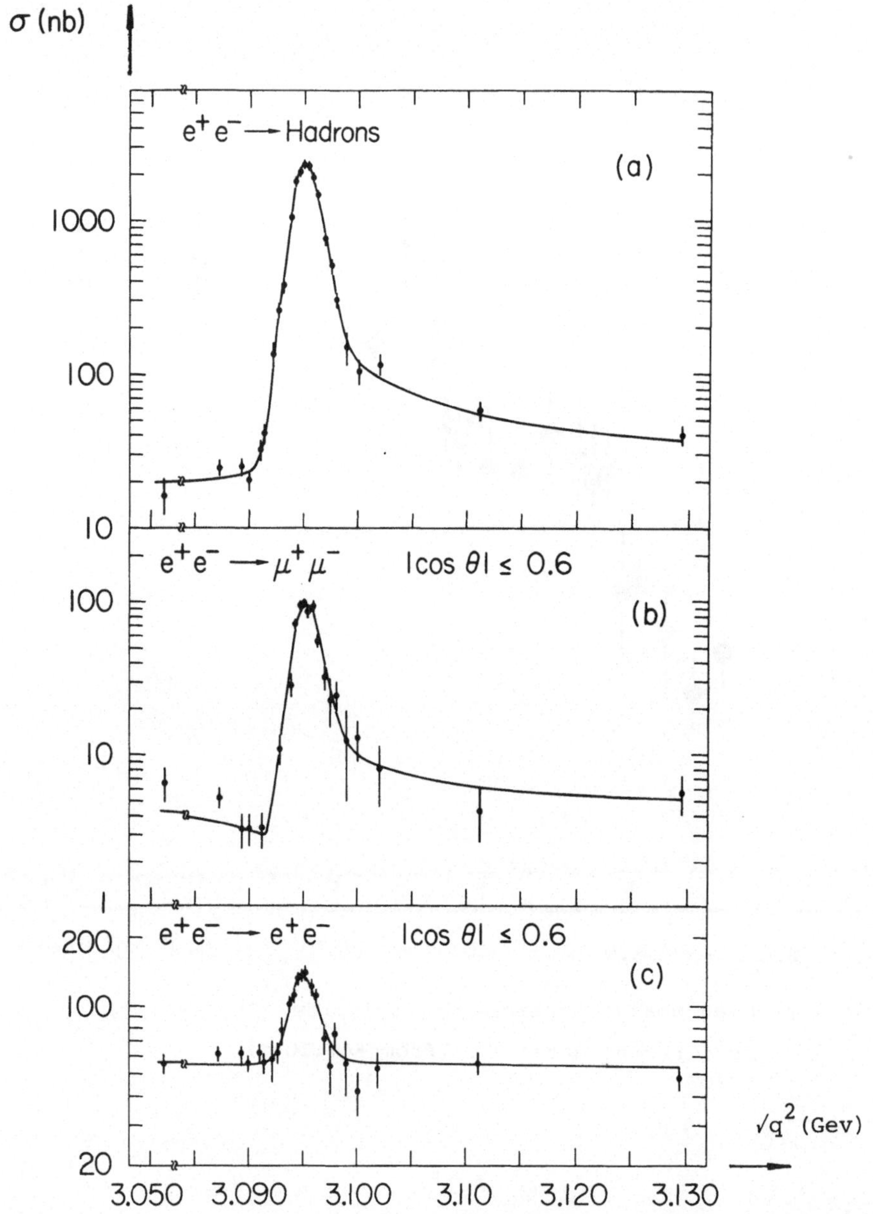

Figure 15 : (a) The total cross section for $e^+e^- \to$ hadrons versus center-of-mass energy

(b), (c) The cross section for $e^+e^- \to \mu^+\mu^-$ and e^+e^-, respectively, versus energy integrated over the range $|\cos\theta| \leq 0.6$ (from Ref.10k)

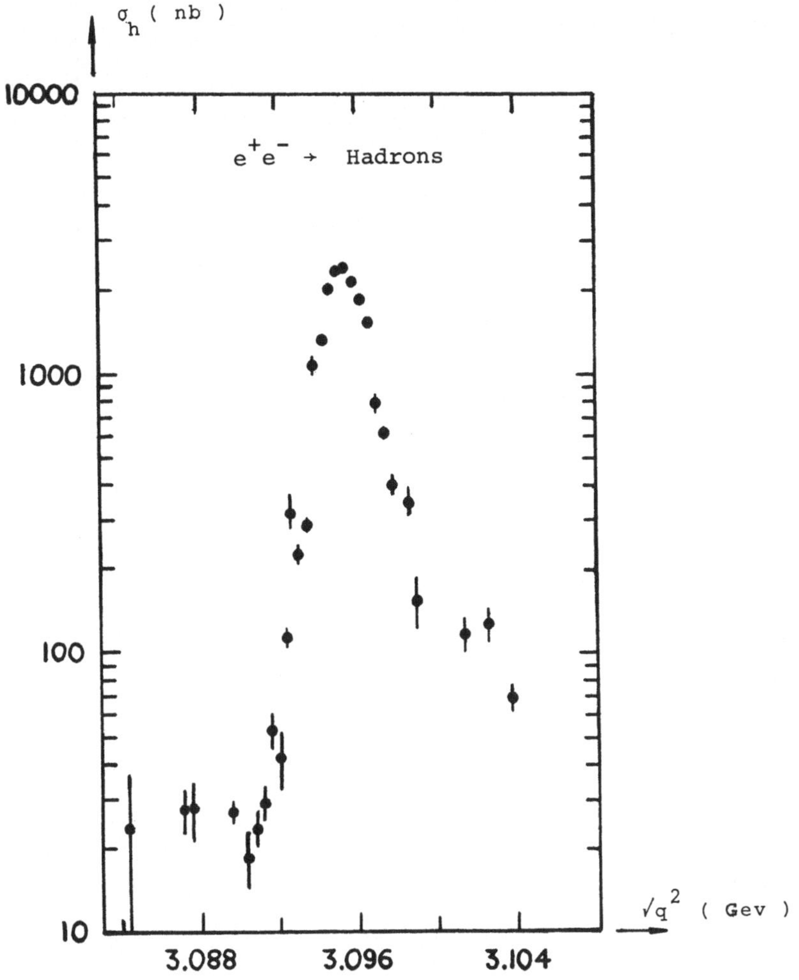

Figure 16 : Total cross section for hadron production vs. center-of-mass energy at $\psi(3.1)$ (from Ref.10i)

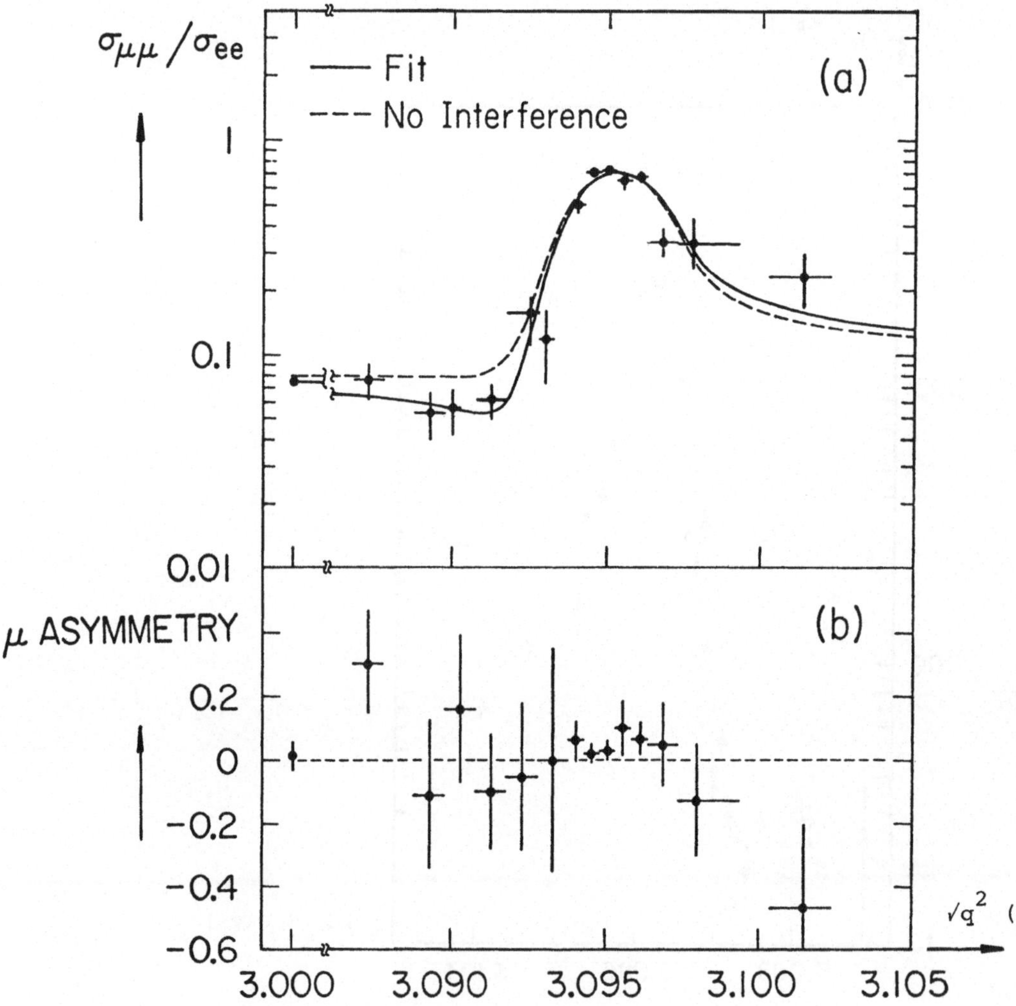

Figure 17 : Interference effect between γ and ψ exchange in $e^+e^- \rightarrow \mu^+\mu^-$ (from Ref. 10k)

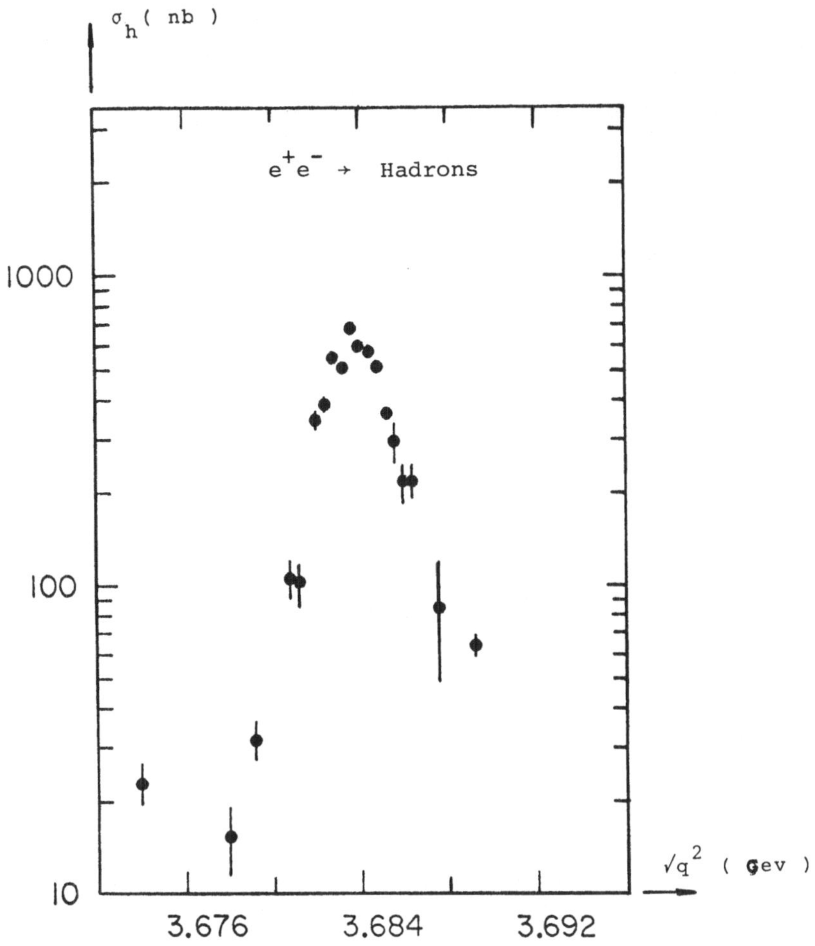

Figure 18 : Total cross section for hadron production vs. center-of-mass energy at $\psi(3.7)$ (from Ref.10i)

EVENTS/0.005 GeV

$\psi'(3.7)\to\pi^+\pi^-+"X"$

(a)

$\psi'(3.7)\to\pi^+\pi^-+"(\iota^+\iota^-)"$

(b)

M_X (GeV)

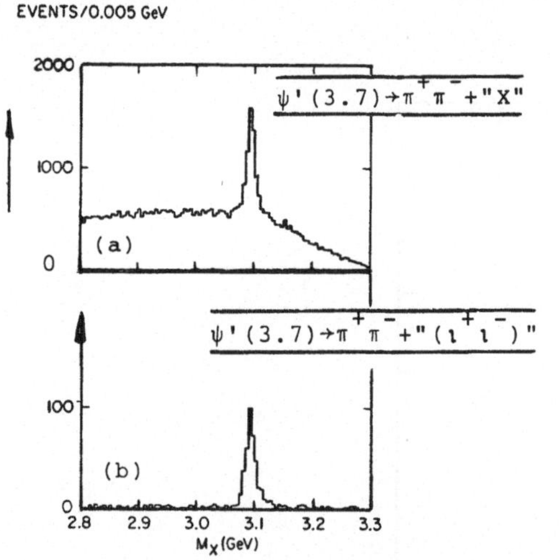

Figure 19a : The distribution of missing mass M_X recoiling against all pairs of oppositly charged particles at the ψ' (from Ref.10f)

Figure 19b : The distribution of the $\mu^+\mu^-$ invariant mass for the highest-momentum oppositely charged particle pair from each ψ' event. Electron pairs are excluded (from Ref.10f)

EVENTS/0.04 GeV

$\psi'(3.7)\to"\mu^+\mu^-"+X$

$M(\mu^+\mu^-)$ (GeV)

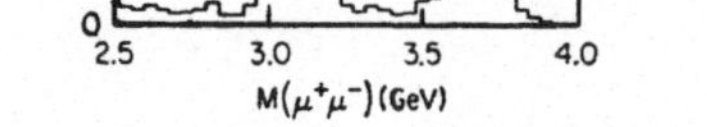

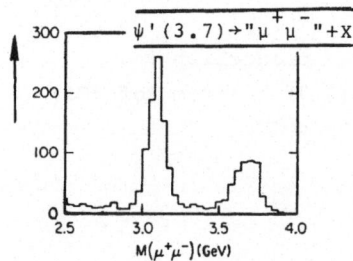

Figure 19c : Distribution of the $\mu^+\mu^-$ effective mass for highest-momentum, oppositely charged particle pairs from each event (from Ref.10f)

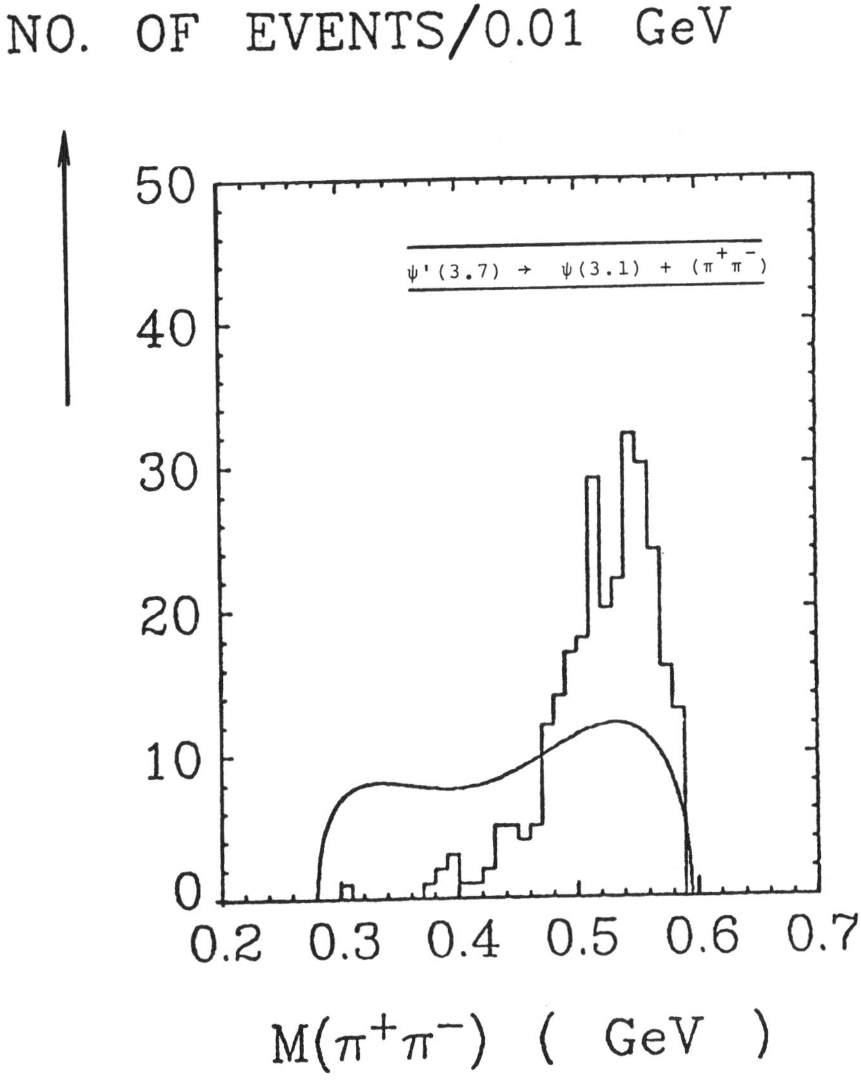

Figure 19d : The distribution of $\pi^+\pi^-$ invariant mass from the decay $\psi' \to \psi + \pi^+\pi^-$. The curve shows the product of phase space times the geometrical acceptance (from Ref.10i)

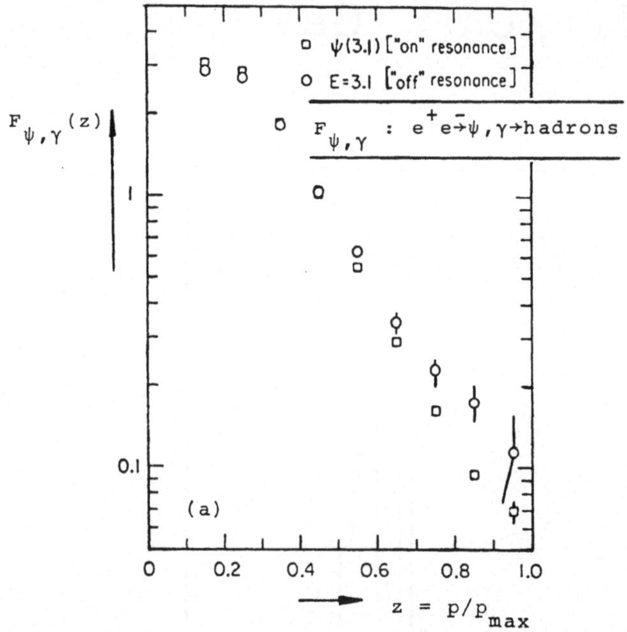

Figure 20a : The normalized momentum distribution for $\sqrt{q^2} = 3.1$ Gev on and off resonance : $F(z) = \frac{1}{\langle n_{ch} \rangle} \cdot \frac{1}{\sigma_h} \cdot \frac{d\sigma_h}{dz}$, $Z = \frac{2p}{\sqrt{q^2}}$ (from Ref.10i)

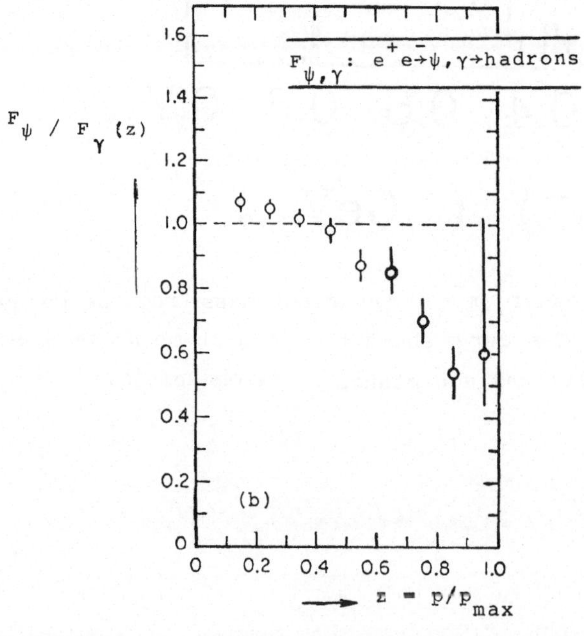

Figure 20b : The ratio of the on and off resonance momentum spectrum (from Ref.10i)

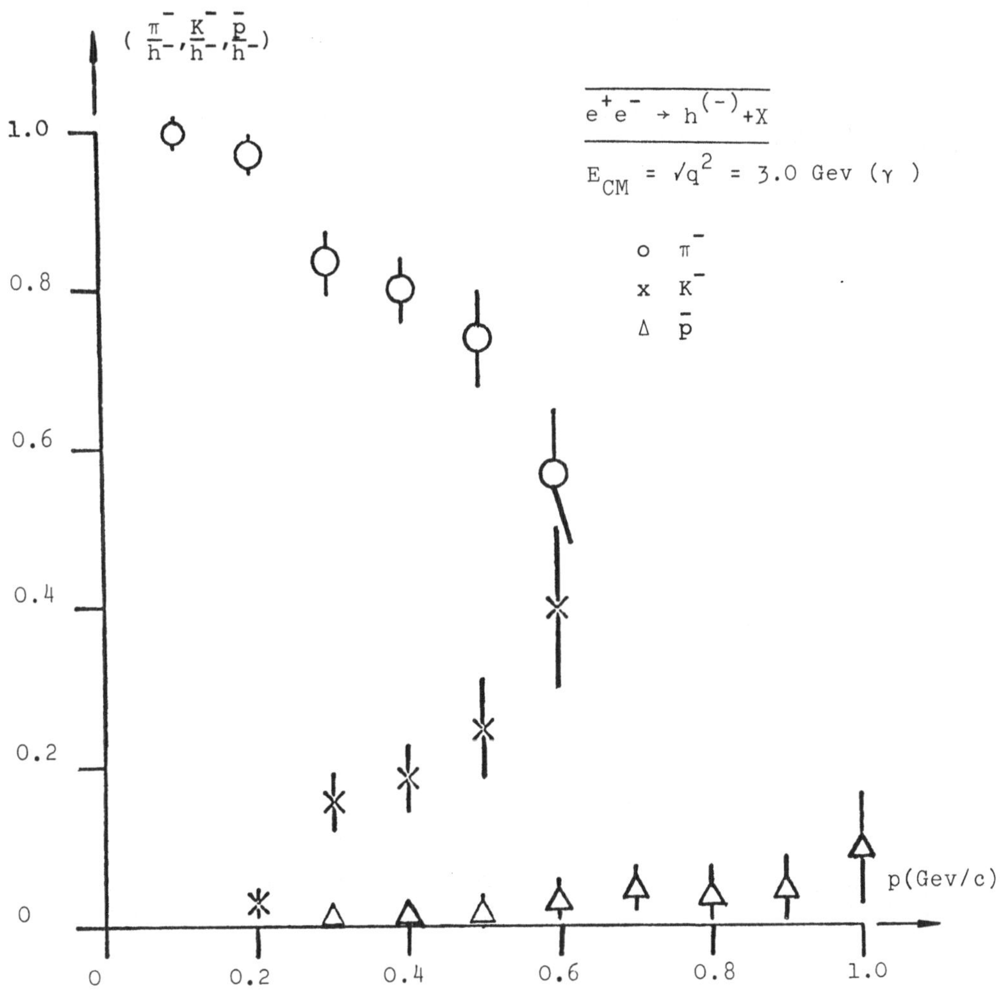

(π⁻/h⁻, K⁻/h⁻, p̄/h⁻)
e⁺e⁻ → h⁽⁻⁾+X
E_CM = √q² = 3.0 Gev (γ)
o π⁻
x K⁻
Δ p̄
1.0
0.8
0.6
0.4
0.2
0
0
0.2
0.4
0.6
0.8
1.0
p(Gev/c)

Figure 21a

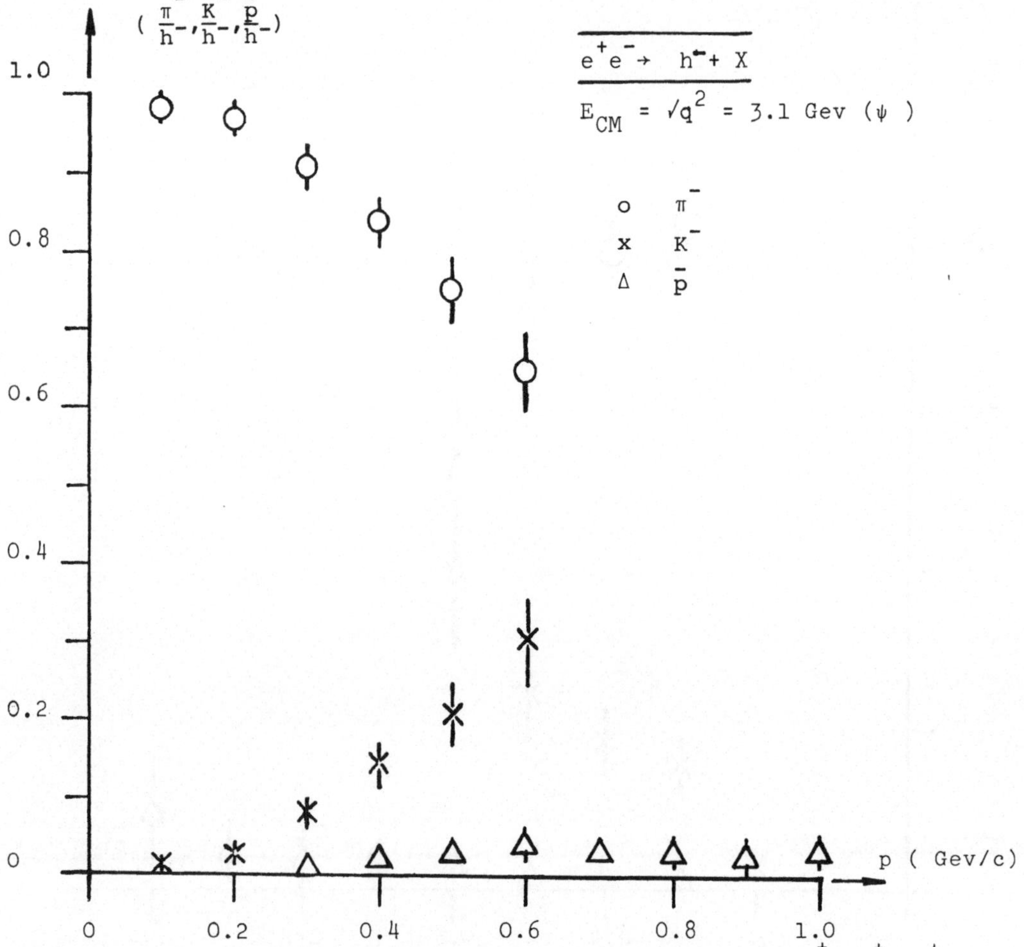

Figure 21a ,b : Inclusive charged particle fractions of $\pi^{\pm}$, $K^{\pm}$, $p^{\pm}$ production at CM-energy $\sqrt{q^2}$ = 3.0 Gev (from Ref.10p)

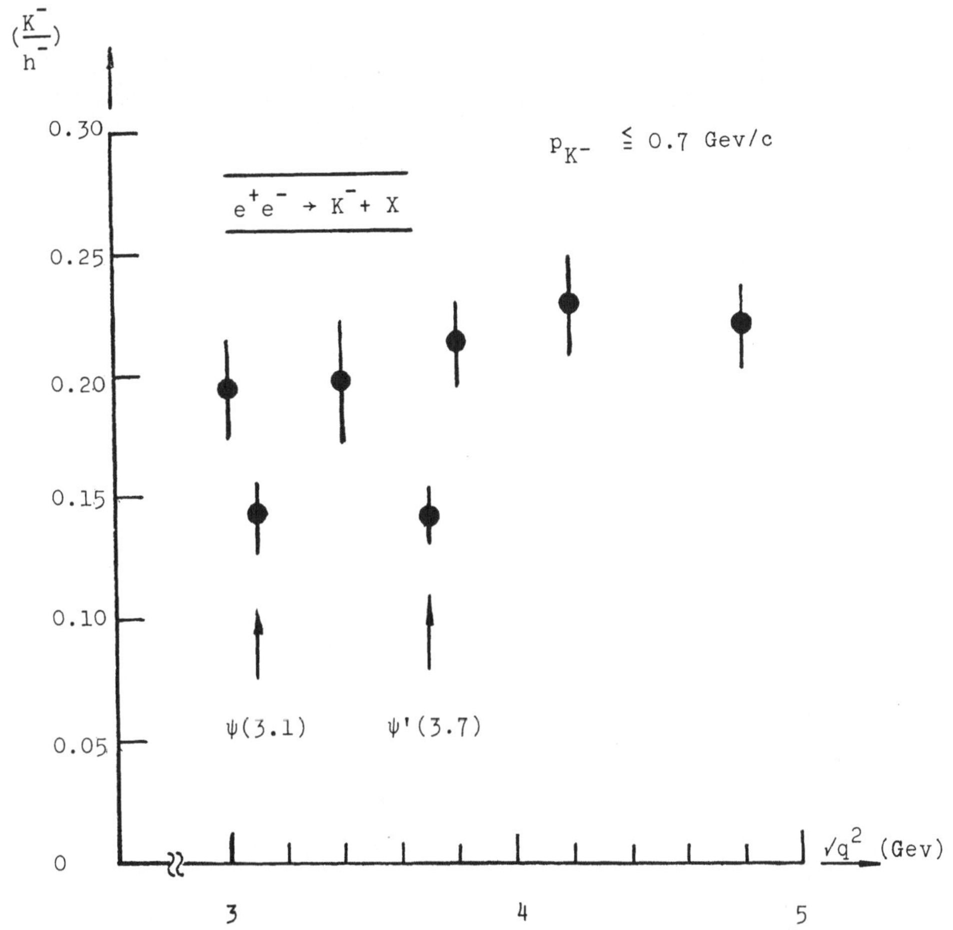

Figure 22 : Fraction of inclusive K^--production $(\frac{K^-}{h^-})$ with momentum $p_{K^-} \lesssim 0.7$ Gev/c as function of the initial energy (from Ref.10 o)

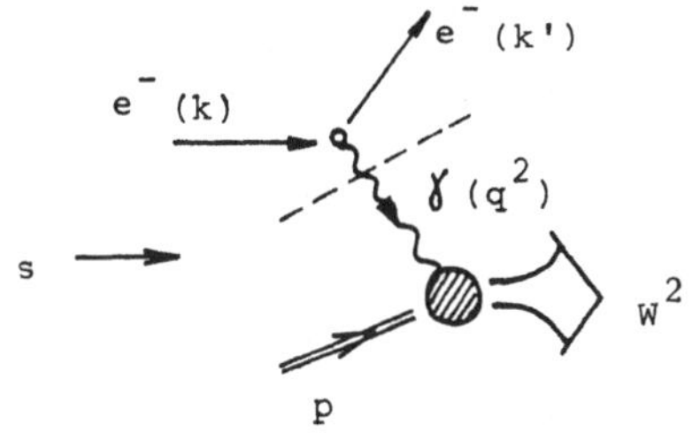

Figure 23 : Deep-inelastic lepton-nucleon scattering $\ell^{\pm} + N \to \ell^{\pm} + X$ with single photon-exchange ($\ell \equiv e, \mu, \nu$)

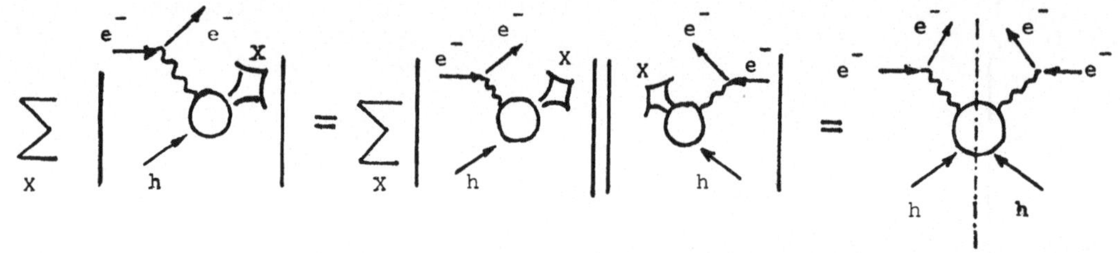

Figure 24 : Graphical representation of the connection between deep inelastic $ep \to eX$ scattering and the off-shell Compton amplitude

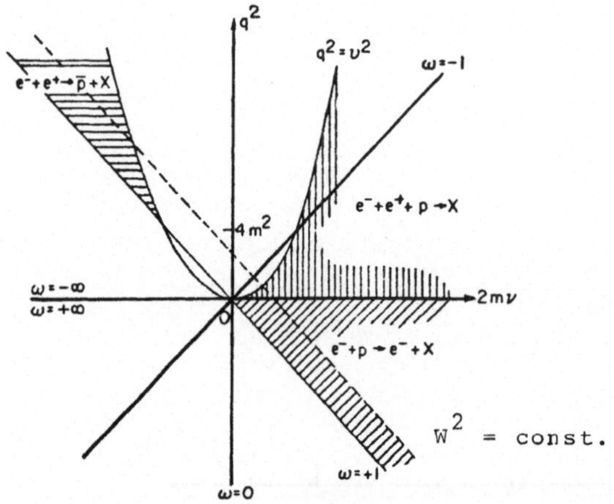

Figure 25 : Physical regions of deep inelastic annihilation : $e^+e^- \to hX$ and deep inelastic scattering: $ep \to eX$ (from Ref.22a)

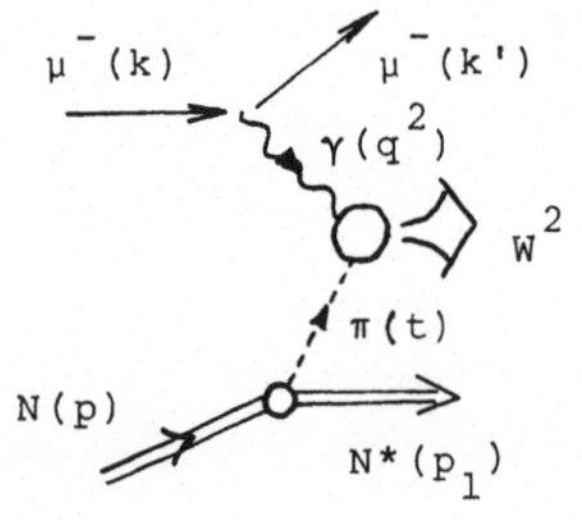

Figure 26 : Deep inelastic $\mu\pi$ -scattering

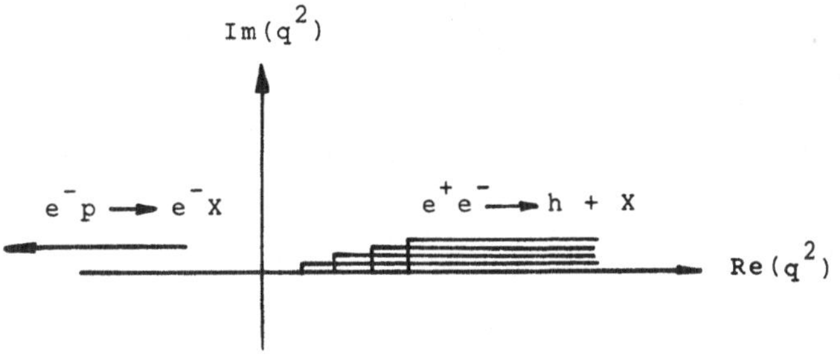

Figure 27 : Complex q^2-plane with production thresholds in e^+e^- annihilation along $q^2 > 0$. Deepinelastic ep scattering for $q^2 < 0$

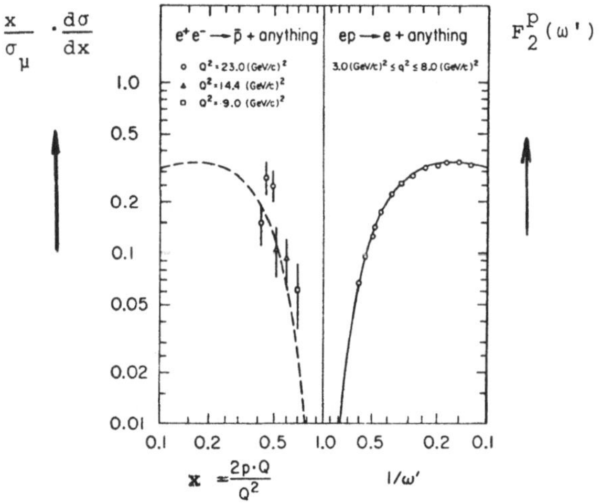

Figure 28 : Values of $\frac{x}{\sigma_\mu} \cdot \frac{d\sigma}{dx}$ for $e^+e^- \to \bar{p} + X$ compared to values of $F_2(\omega')$ for $ep \to e + X$. The solid line is a fit to the electroproduction data and its "reflection", the dashed line, is what is expected from eq. (3.23) (see text) (from Ref.3a)

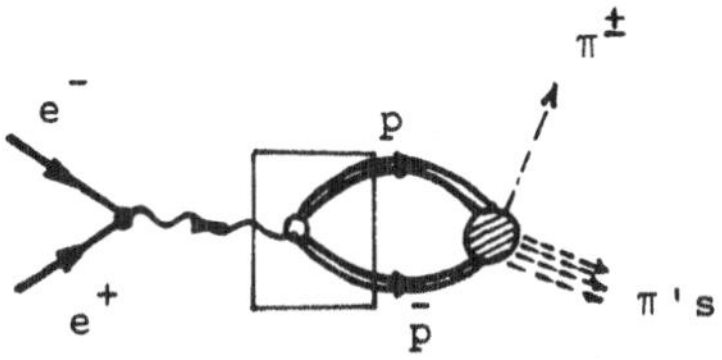

Figure 29 : π-production in e^+e^- annihilation via inclusive $p\bar{p}$ intermediate states

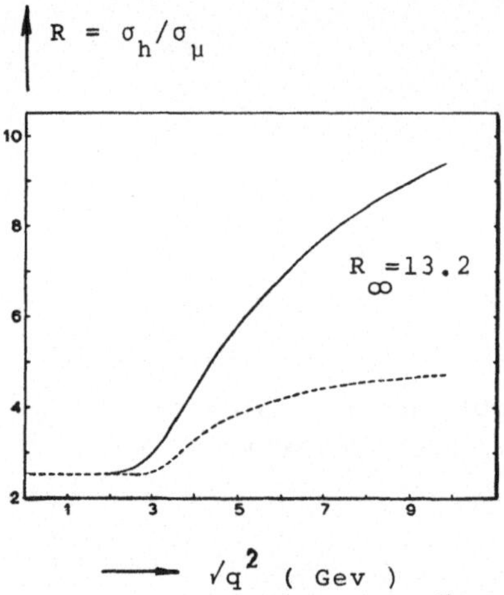

Figure 30a : $R(q^2) \equiv \sigma_h/\sigma_\mu$ in the model of DiGiacomo-Konishi (Ref. 24)

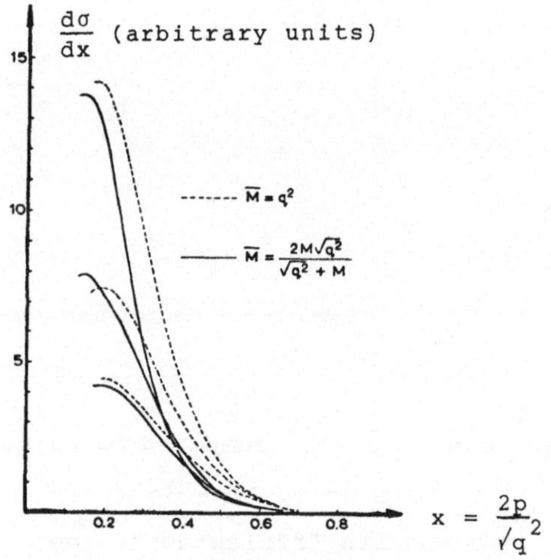

Figure 30b : Inclusive distribution in the model of DiGiacomo-Konishi (Ref. 24)

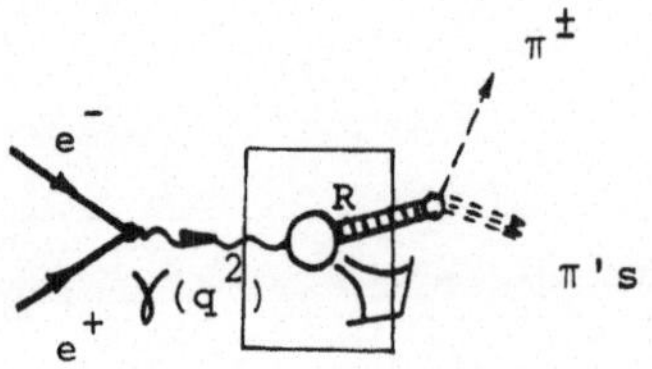

Figure 31 : π-production in e^+e^- annihilation via inclusive resonance intermediate states

$R = \sigma_h / \sigma_\mu$

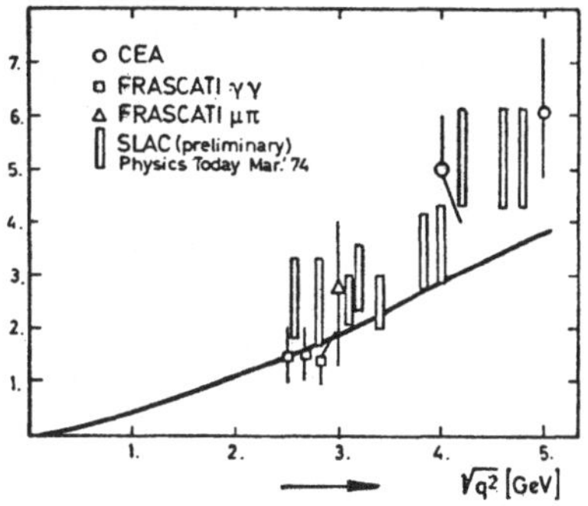

Figure 32a : $R(q^2) \equiv {}^{\sigma_h}/_{\sigma_\mu}$ in the model of Schierholz- Schmidt (Ref. 25)

$q^2 \frac{d\sigma}{dx}(\pi^\pm)$ $(\mu b \cdot Gev^2)$

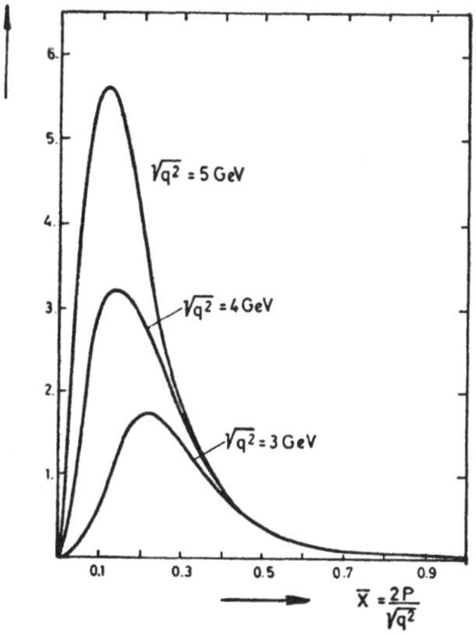

Figure 32b : Inclusive distribution in the model of Schierholz-Schmidt (Ref. 25)

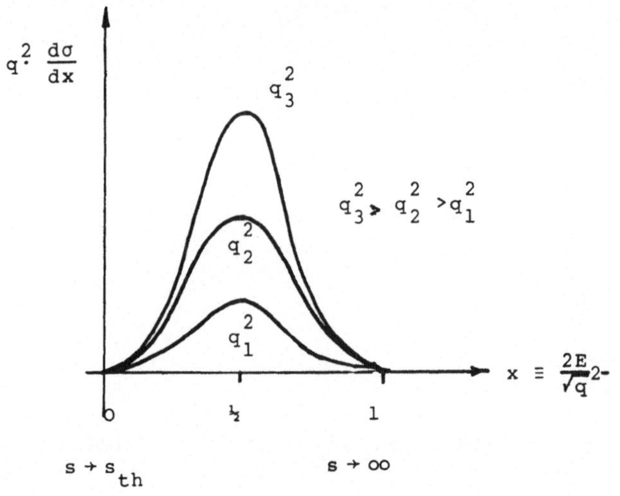

Figure 33 : $e^+e^- \to \bar{\pi} + R$: Inclusive distribution $q^2 \frac{d\sigma}{dx}$ vs. x ; it grows with increasing q^2 for x = fixed

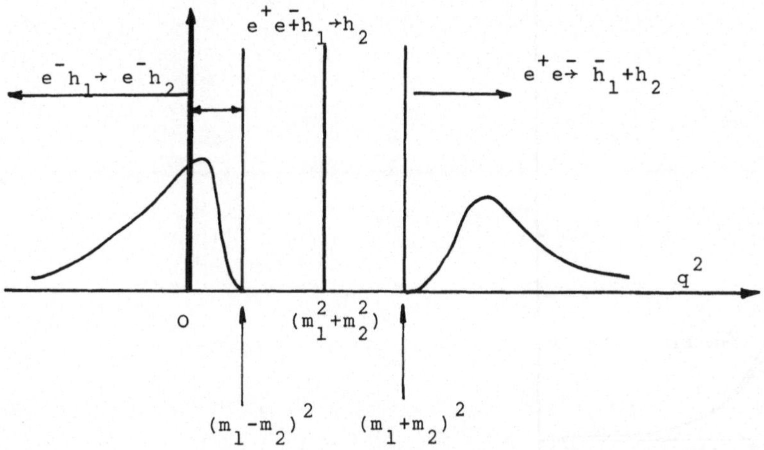

Figure 34 : The qualitative behaviour as a function of q^2 to be expected of the cross sections for $\gamma(q^2) + \bar{h}_2 \to h_1$ and $\gamma(q^2) \to h_1 + h_2$

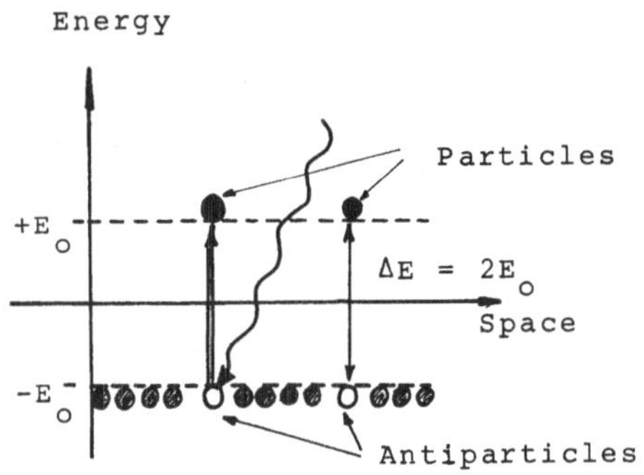

Figure 35 : Dirac's interpretation of a completely filled negative energy spectrum : Dirac-sea

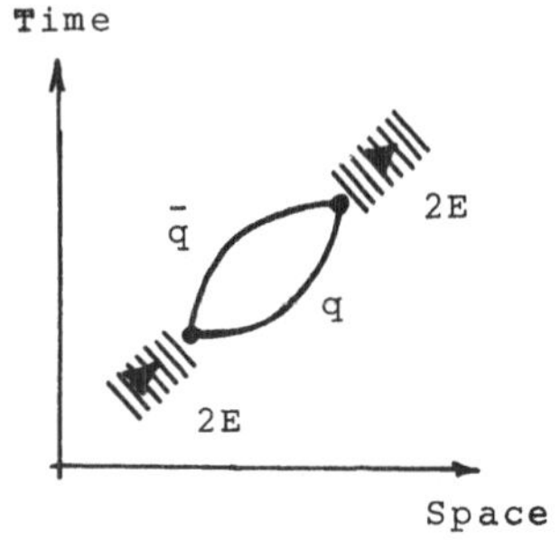

Figure 36a : Process of particle creation in Dirac's vacuum. Particle-antiparticle creation

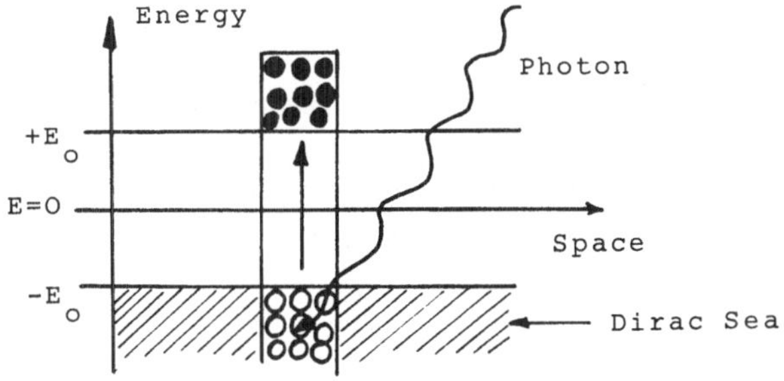

Figure 36b : Process of particle creation in Dirac's vacuum. Many particle production (→ Fireball)

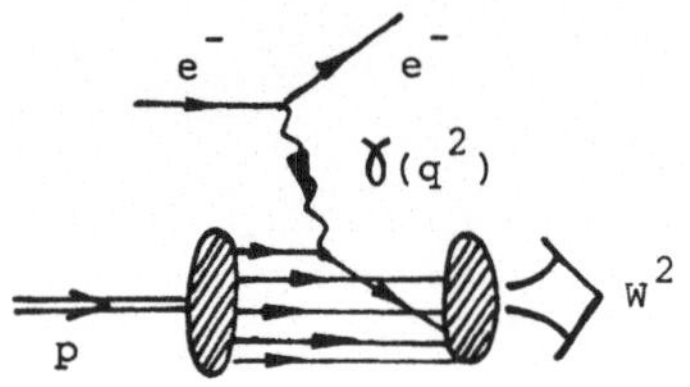

Figure 37 : Parton model for deep inelastic ep scattering

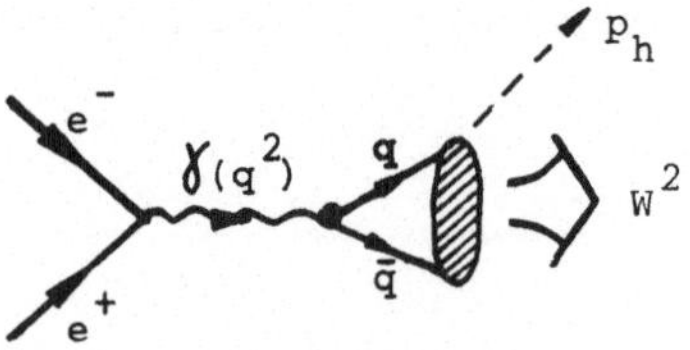

Fig.38a

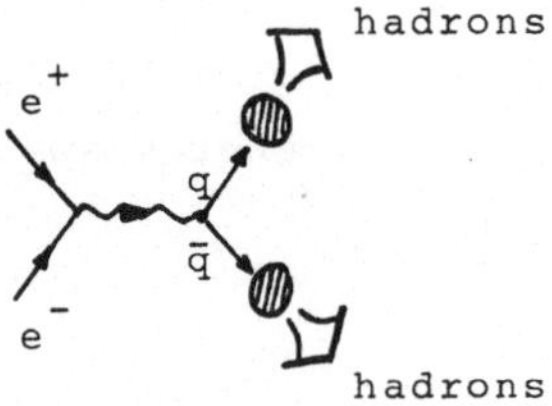

Fig.38b

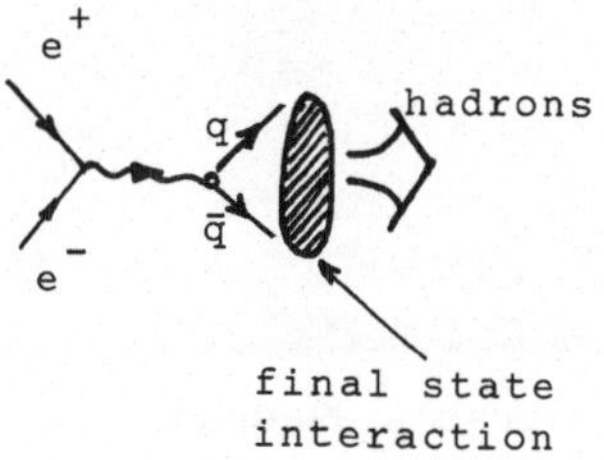

Fig.38c

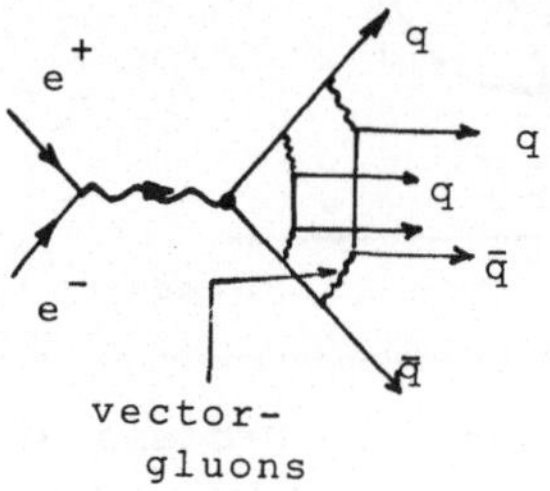

Fig.38d

Figure 38 : Parton model

(a) Inclusive hadron production

(b) Hadron production without $q\bar{q}$-interactions

(c) Hadron production with $q\bar{q}$-interactions (Fan-model, Final-state interactions, etc.)

(d) Hadron production process viewed as a screening process

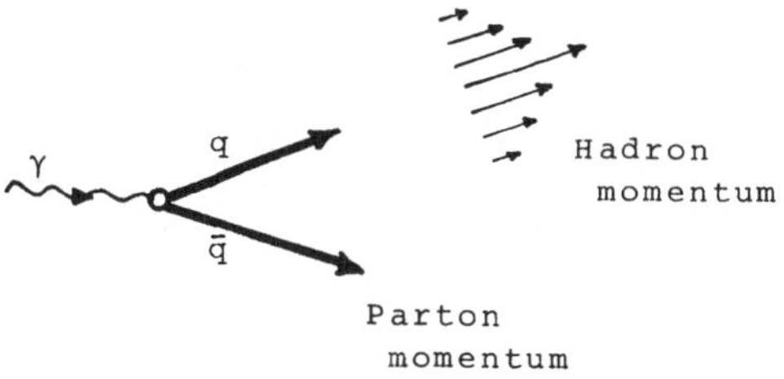

Figure 39 : Parton-hadron transition defined by hadron- momentum distribution function $D_i(x)$

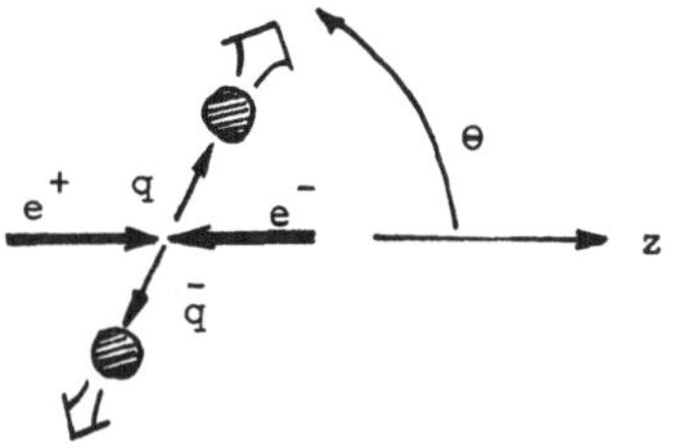

Figure 40 : Jet structure in the parton model for e^+e^- annihilation

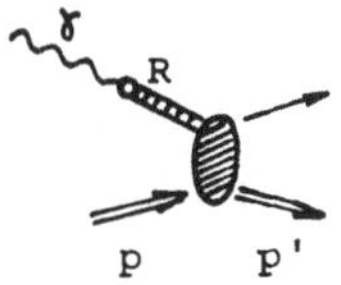

Fig.41

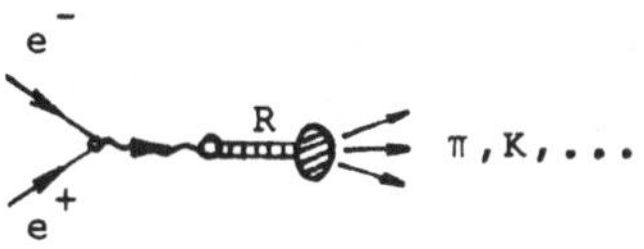

Fig.42

Figure 41 : Vector meson dominance (VMD) in (quasi) two- particle photoproduction processes

Figure 42 : VMD in e^+e^- annihilation : $R \equiv \{\rho,\omega,\phi\}$

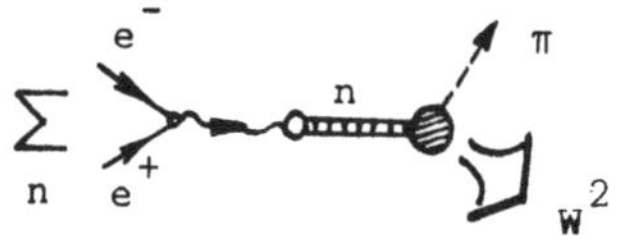

Figure 43 : Extended vector meson dominance (EVMD) : $R \equiv n,\ m_n^2 = m_\rho^2 \cdot (1 + 2\ n)$

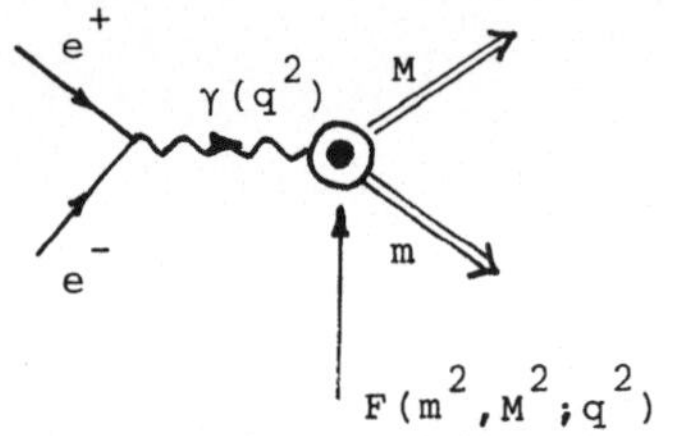

Figure 44 : Formfactor and two-particle threshold factor in the cut-model

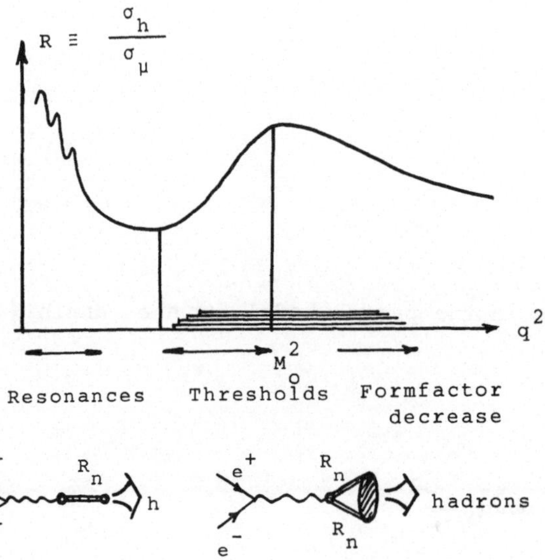

Figure 45 : Cross section behaviour in the cut-model

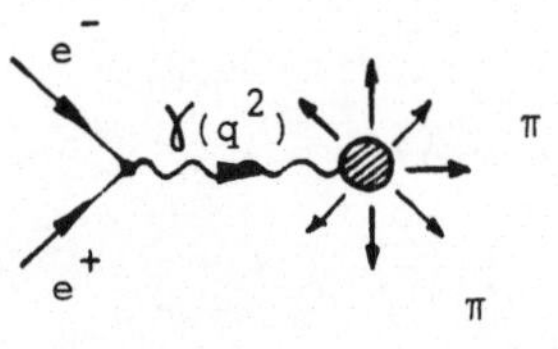

Figure 46 : Hadron production in Fermi's statistical model (FM)

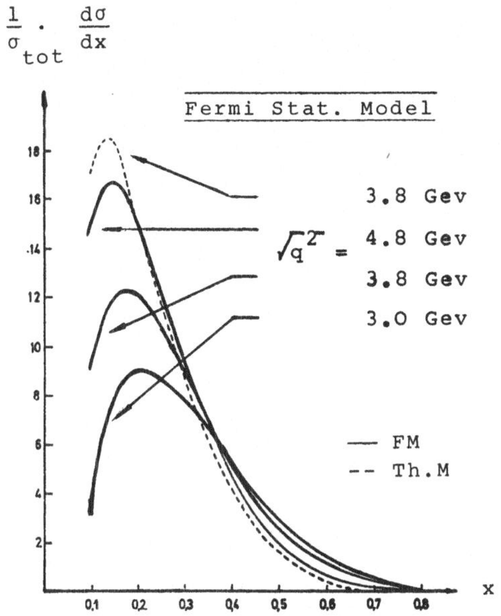

Figure 47 : Inclusive distribution $\frac{1}{\sigma_{tot}} \cdot \frac{d\sigma}{dx}$ vs. x in FM.
The dotted curve is from the thermodynamic (TM) model at $\sqrt{q^2}$ = 3.8 Gev with Hagedorn temperature $k\,T_o$ = 193 Mev (from Ref.47)

Figure 48a,b : Hadron production in e^+e^- annihilation described by Landau's hydrodynamical model (LM)

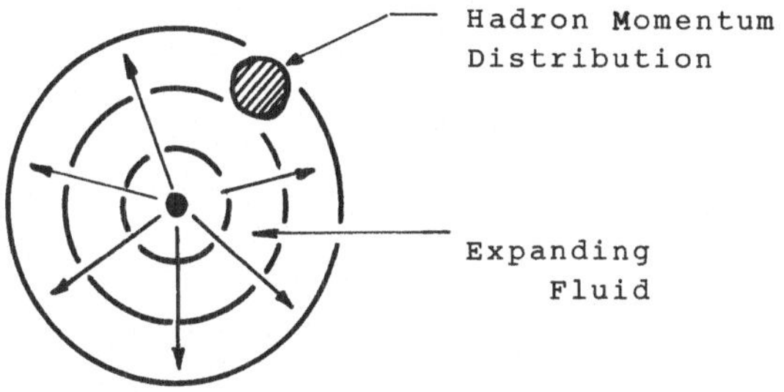

Figure 48b

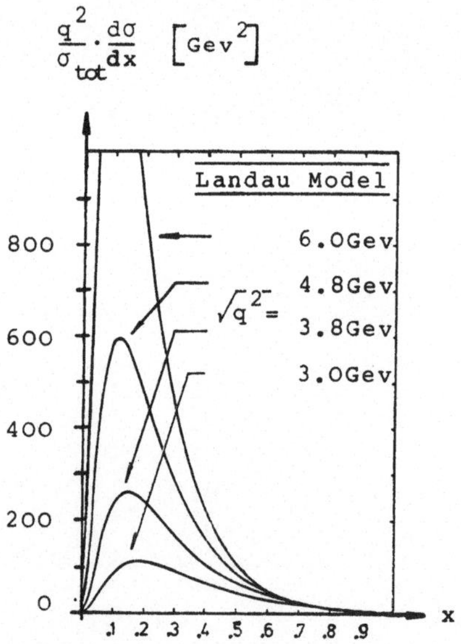

Figure 49 : Inclusive distribution $\frac{q^2}{\sigma_{tot}} \cdot \frac{d\sigma}{dx}$ vs. x in Landau's hydrodynamical model (LM) (from Ref.51g)

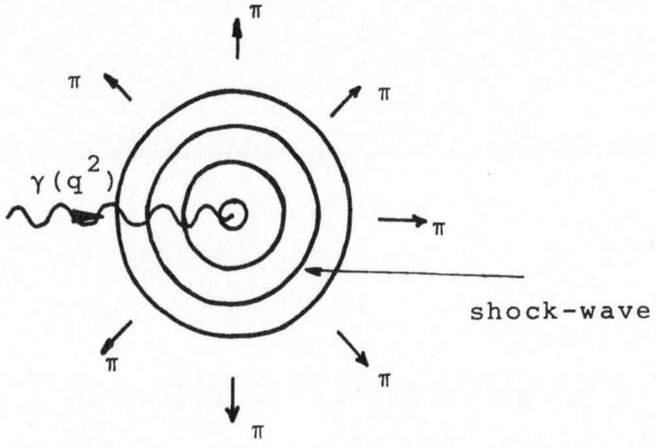

Figure 50 : Hadron production in Heisenberg's statistical model (HM)

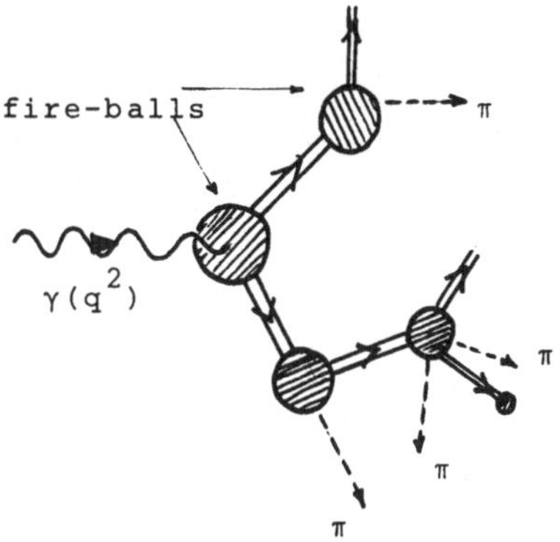

Figure 51 : Hadron production in e^+e^- annihilation described by the thermodynamical model (TM)

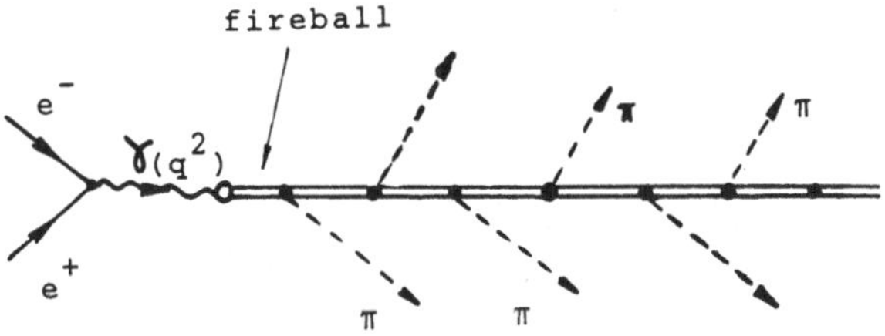

Figure 52 : Hadron production in e^+e^- annihilation described by chain-decay models (CM)

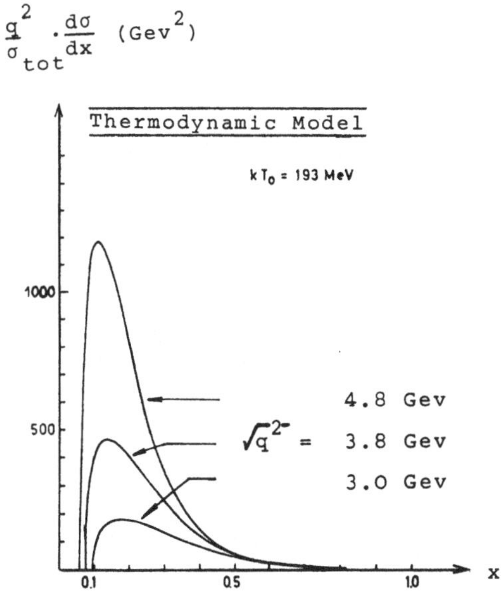

Figure 53 : Inclusive distribution $\frac{q^2}{\sigma_{tot}} \cdot \frac{d\sigma}{dx}$ vs. x in the thermodynamical model (TM) (from Ref.56a)

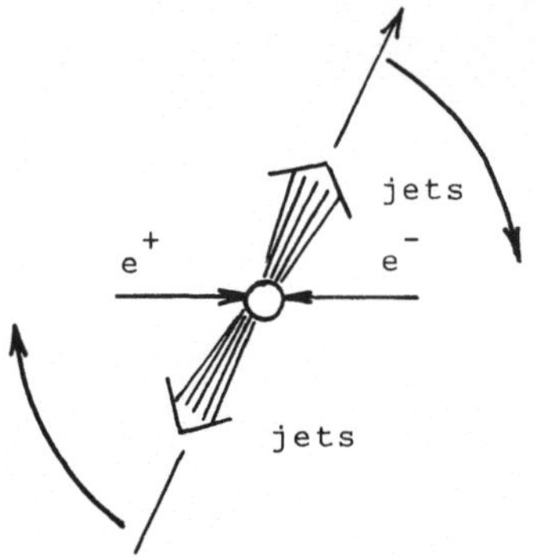

Figure 54 : Hadron production in e^+e^- annihilation in the uncorrelated jet-model (JM)

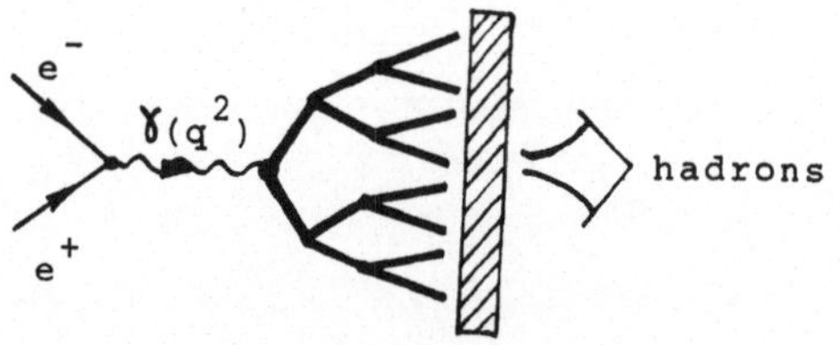

Fig.56a

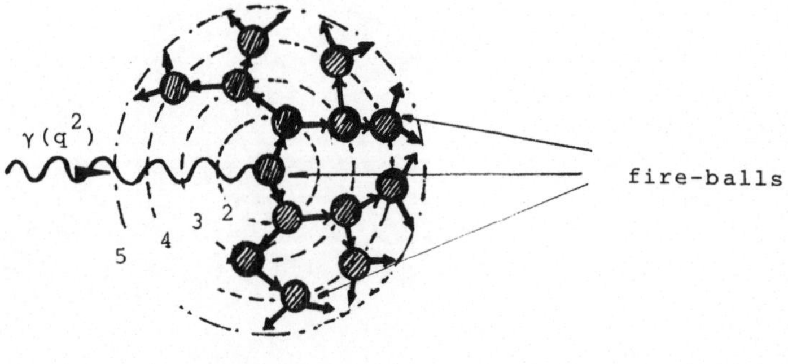

Fig.56b

Figure 56a,b : Hadron production in e^+e^- annihilation as viewed in the cascade decay model

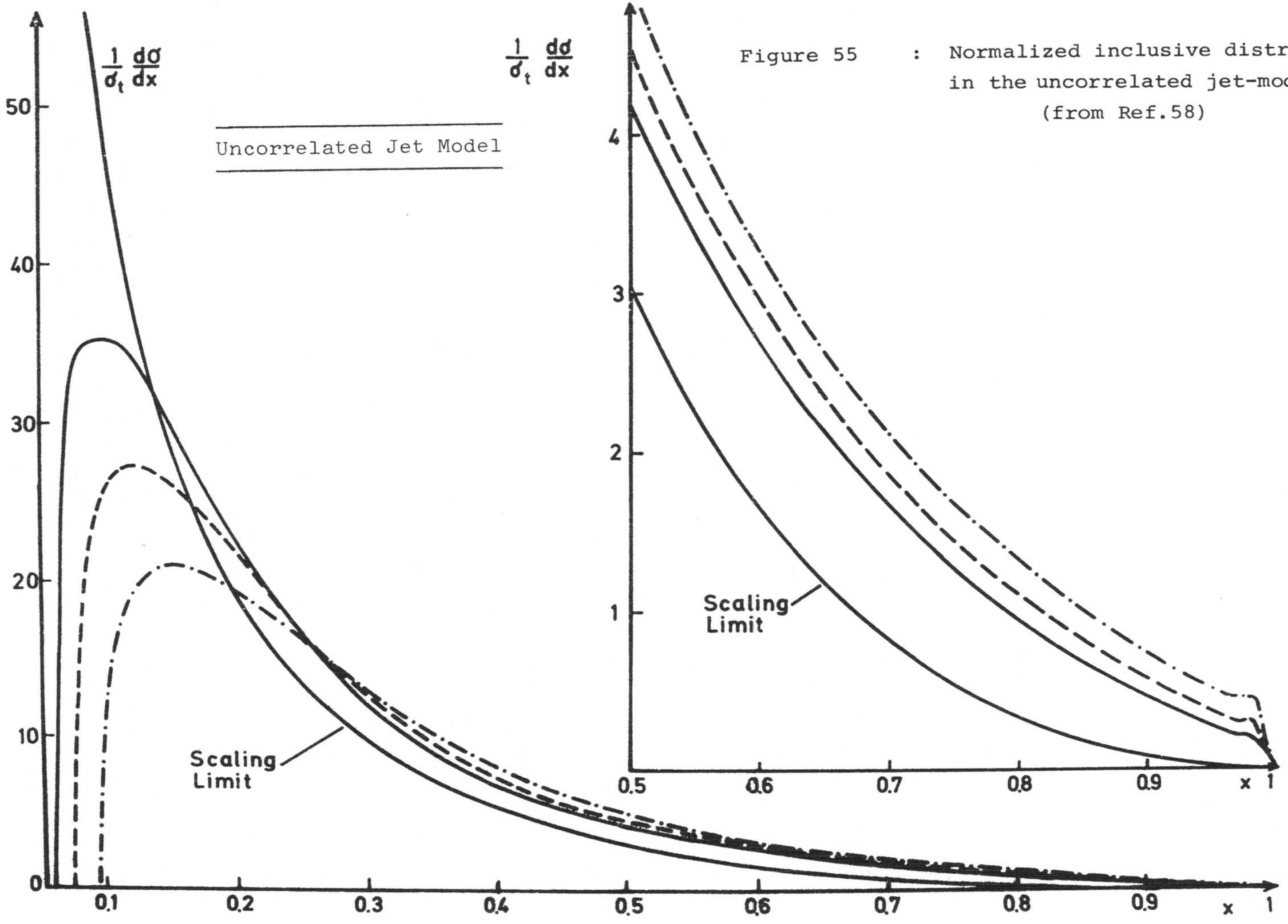

Figure 55 : Normalized inclusive distribution in the uncorrelated jet-model (JM) (from Ref.58)

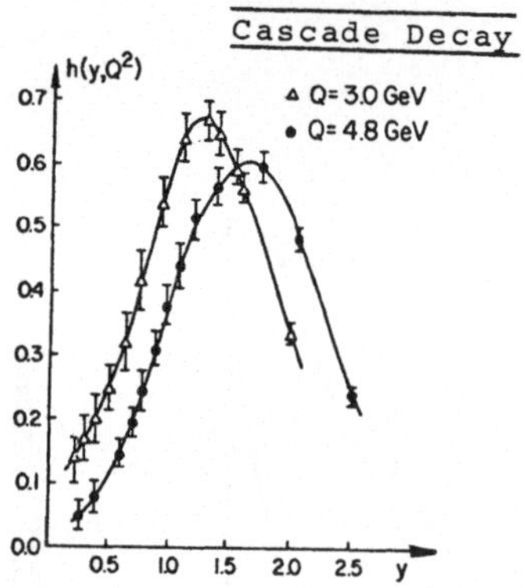

Figure 57 : Inclusive distribution in the cascade decay model (from Ref. 59c)

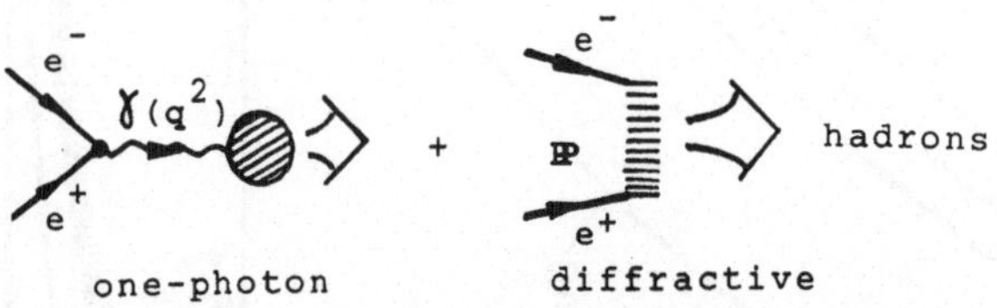

Figure 58 : Hadron production in e^+e^- annihilation via single photon - and P-exchange ("Pomeron")

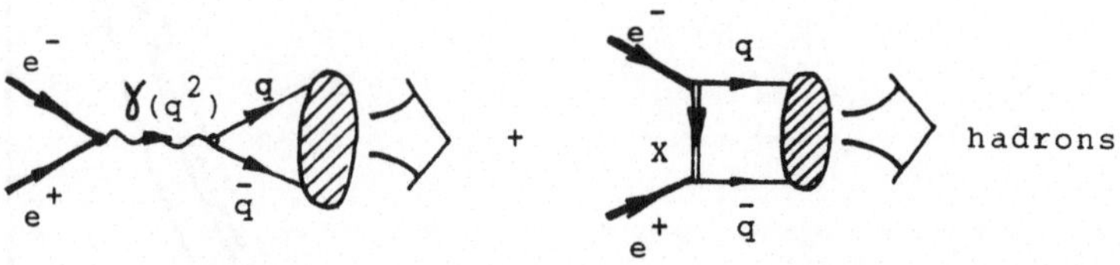

Figure 59 : Hadron production in e^+e^- annihilation via single photon - and lepto-hadron X-exchange

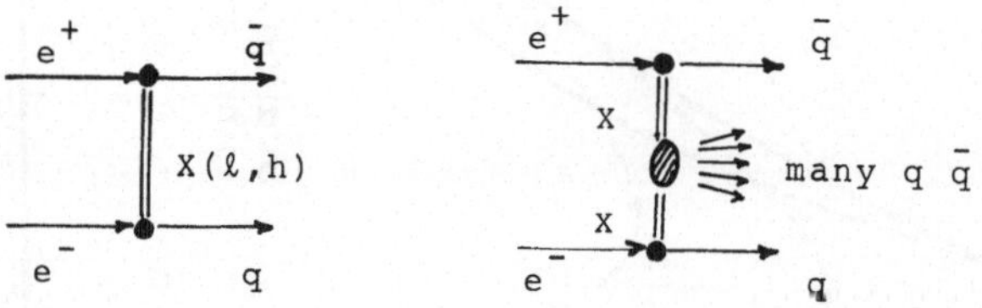

Figure 60 : Multiperipheral characteristics in a new interaction model with lepto-hadron exchange X(ℓ, h)

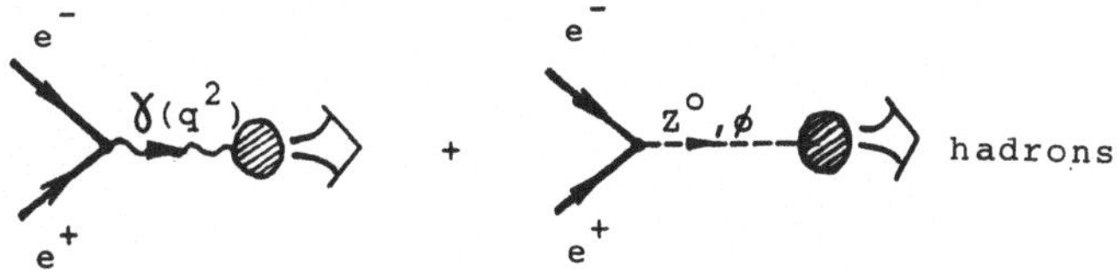

Figure 61 : Hadron production in e^+e^- annihilation via single photon - and gauge boson Z^O - or Higgs scalar meson exchange

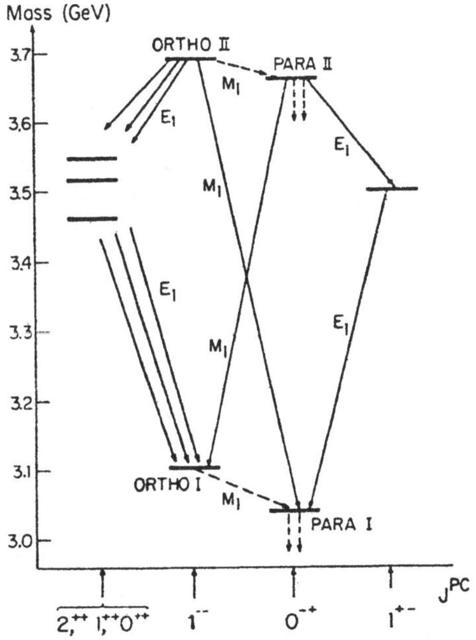

Figure 62 a : Masses and radiative transitions of Charmonium (from Ref.109a)

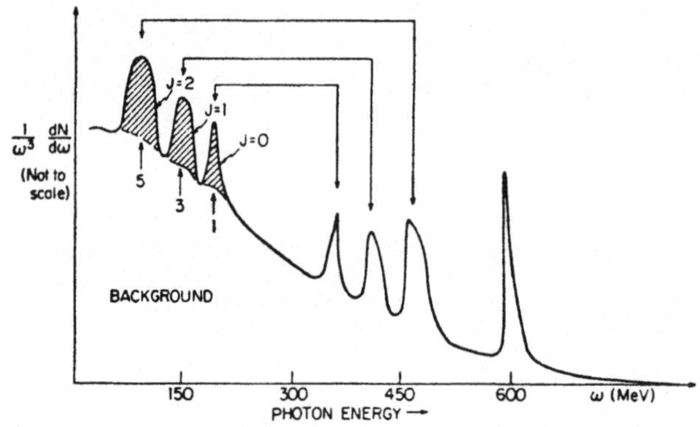

Figure 62 b : Expected photon spectrum due to radiative transitions between different Charmonium states

Ortho II → $^3P_J + \gamma$: first three bumps

^3P_J → Ortho I + γ : second three bumps

Ortho II → Para I+γ : last bump (from Ref.110)

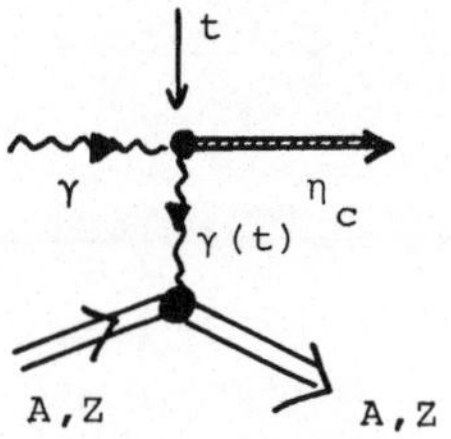

Figure 63 : Primakoff photoproduction of η_c

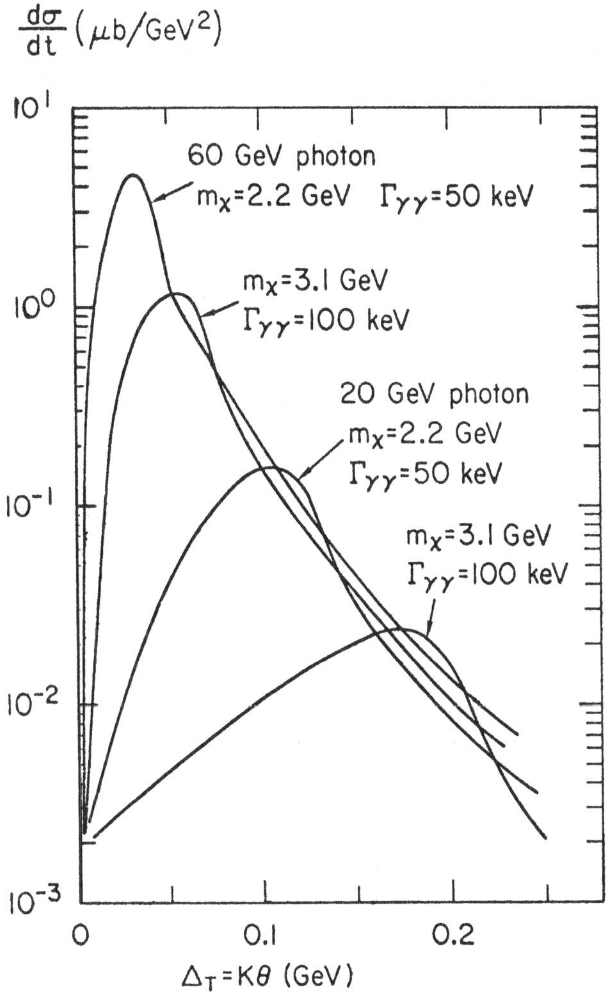

Figure 64 : Differential cross section of Primakoff photoproduction of pseudoscalar η_c of Pb (from Ref.73d)

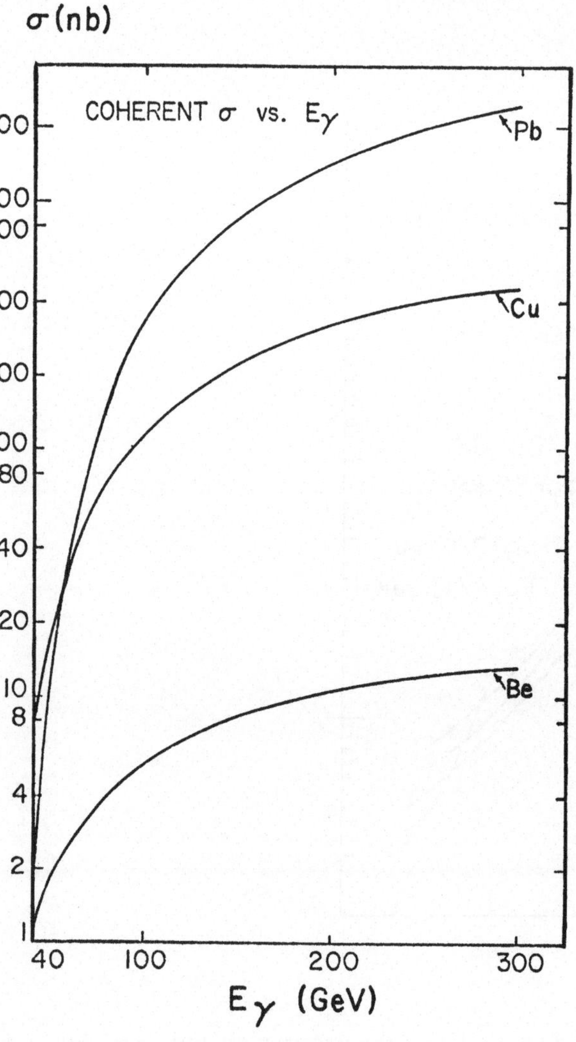

Figure 65 : Energy dependence of the total integrated η_c-production cross section for various nuclei (from Ref.73a)

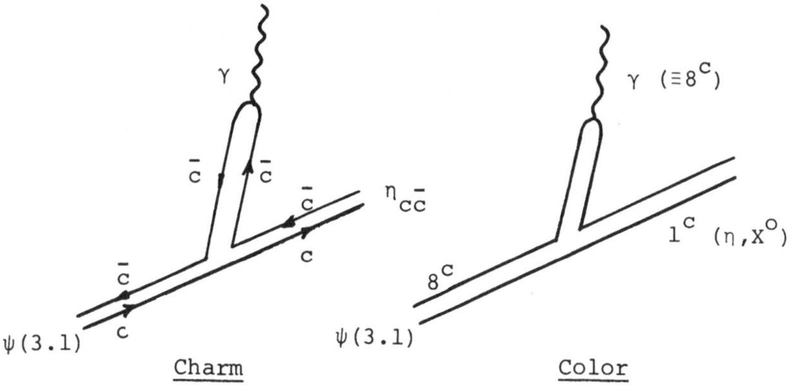

Figure 66 : Photon emission in the charm and color cases

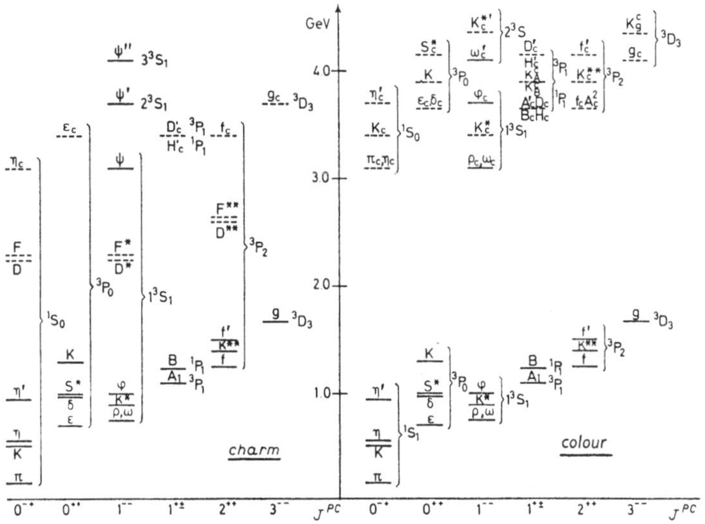

Figure 67 : Levels in the charm and color schemes

Charm: Quadratic mass formula for vector states plus degeneracy of L-levels in the quark model

Color: $\psi(3.1) \equiv \omega_c$ and ψ (3.7) $\equiv$ ϕ_c and nonet symmetry for the other states with ε_c, δ_c, η_c pure $u\bar{u}+d\bar{d}$

(from Ref.85d)

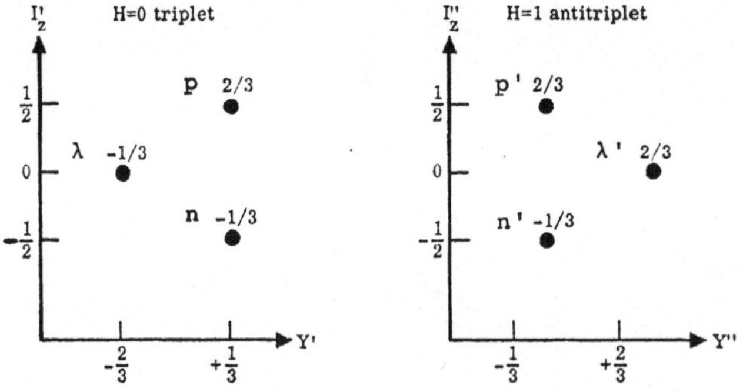

Figure 68 : Quantum number assignment in the heavy quark model (Ref. 93)

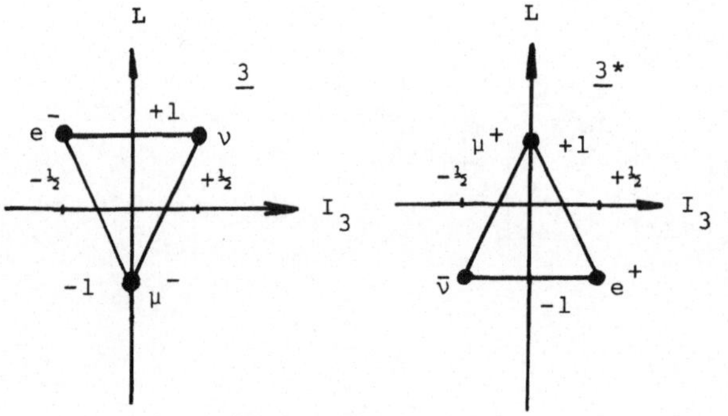

Fig. 69

Figure 69 : Quantum number assignment to leptons (Ref. 97)

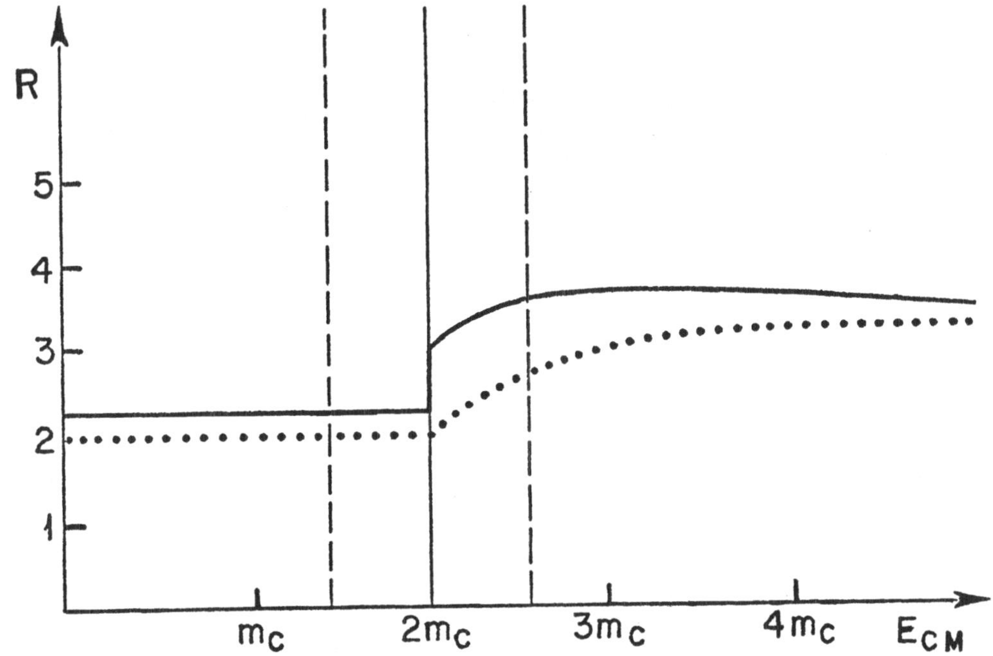

Figure 70 : Step increase in R as viewed in the Charmonium model (from Ref.106b)

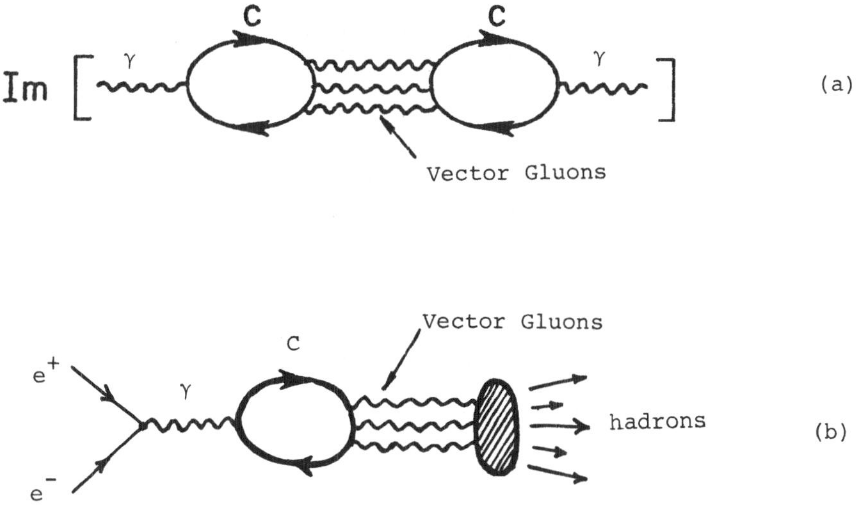

Figure 71a,b : Hadron production in the Charmonium model via gluon exchange

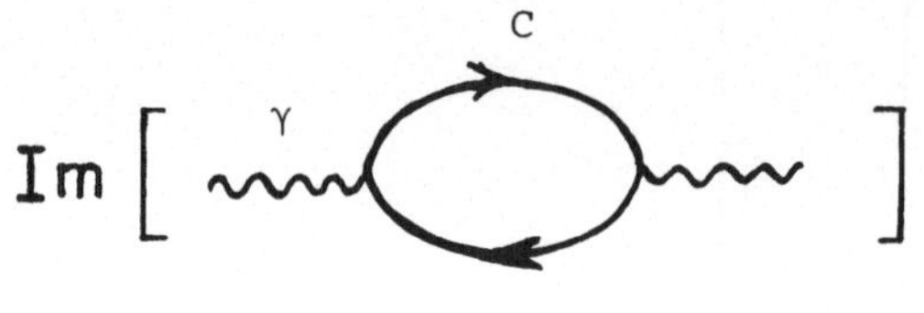

(a)

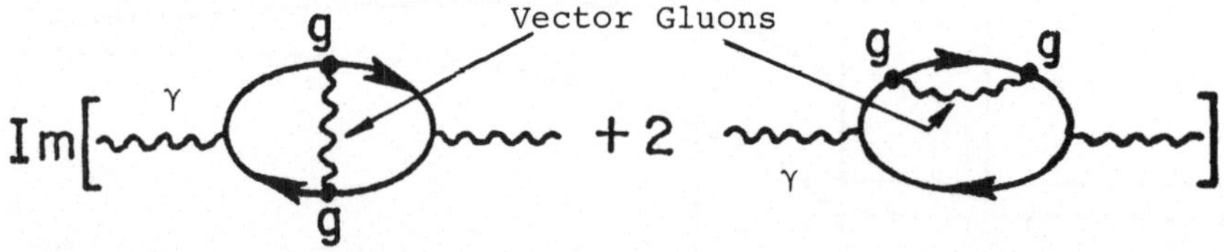

(b)

Figure 72a,b : Higher order corrections in the Charmonium model due to multi-gluon exchange

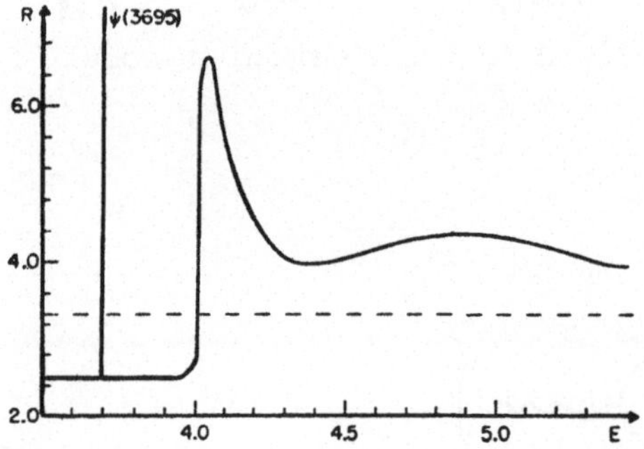

Figure 73 : Increase of $R \equiv \sigma_h/\sigma_\mu$ near threshold if a virtual s-wave bound state is assumed about 95 Mev above threshold at 4.0 Gev (from Ref.111a)

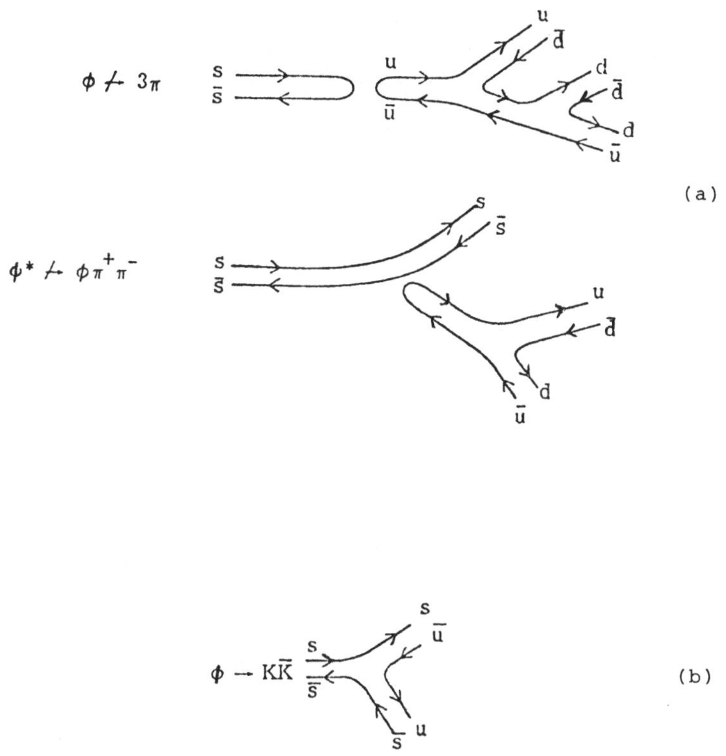

Figure 74a,b : Graphical representation of Zweig's selection rule for typically forbidden decays as $\phi \not\to 3\pi$ or $\phi^* \not\to \phi + \pi\pi$ and for an allowed decay $\phi \to K\bar{K}$

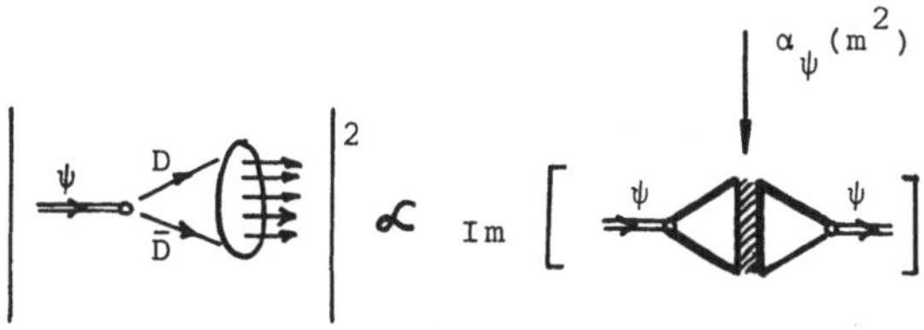

Figure 75 : Two step decay : $\psi \to D\bar{D} \to$ hadrons

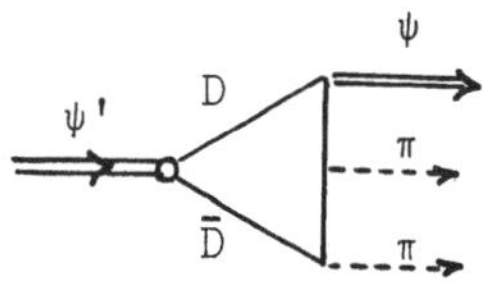

Figure 76 : $\psi' \to \psi + \pi\pi$ decay in the two-step-model

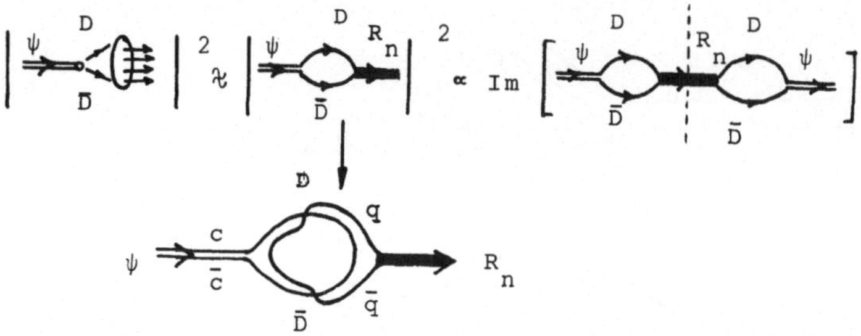

Figure 77 : Feynman diagram for $\psi \rightarrow D\bar{D} \rightarrow$ " hadrons" and corresponding double twist diagram

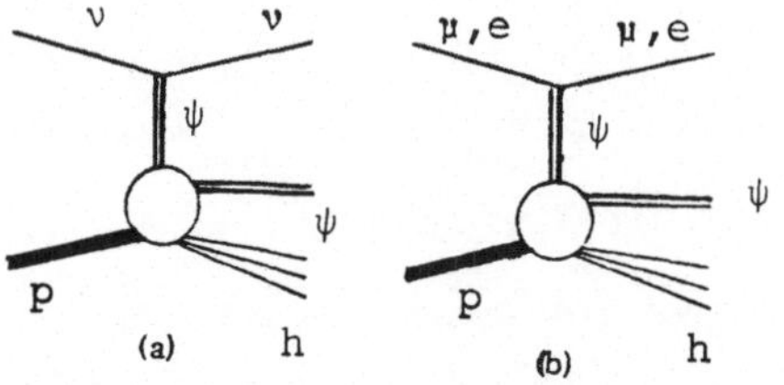

Figure 79 a,b: Deep inelastic ψ-production if a ψ-h-ψ coupling is supposed to exist

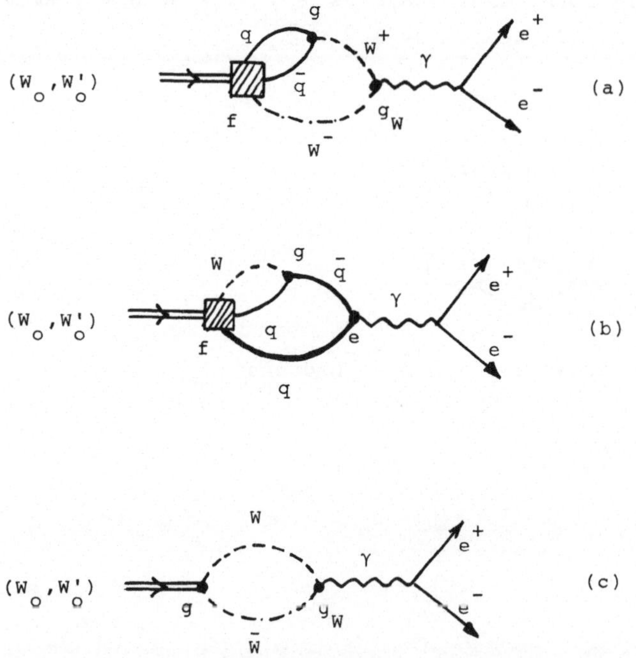

Figure 80a,b,c: Lepton production in the strong W-pair model

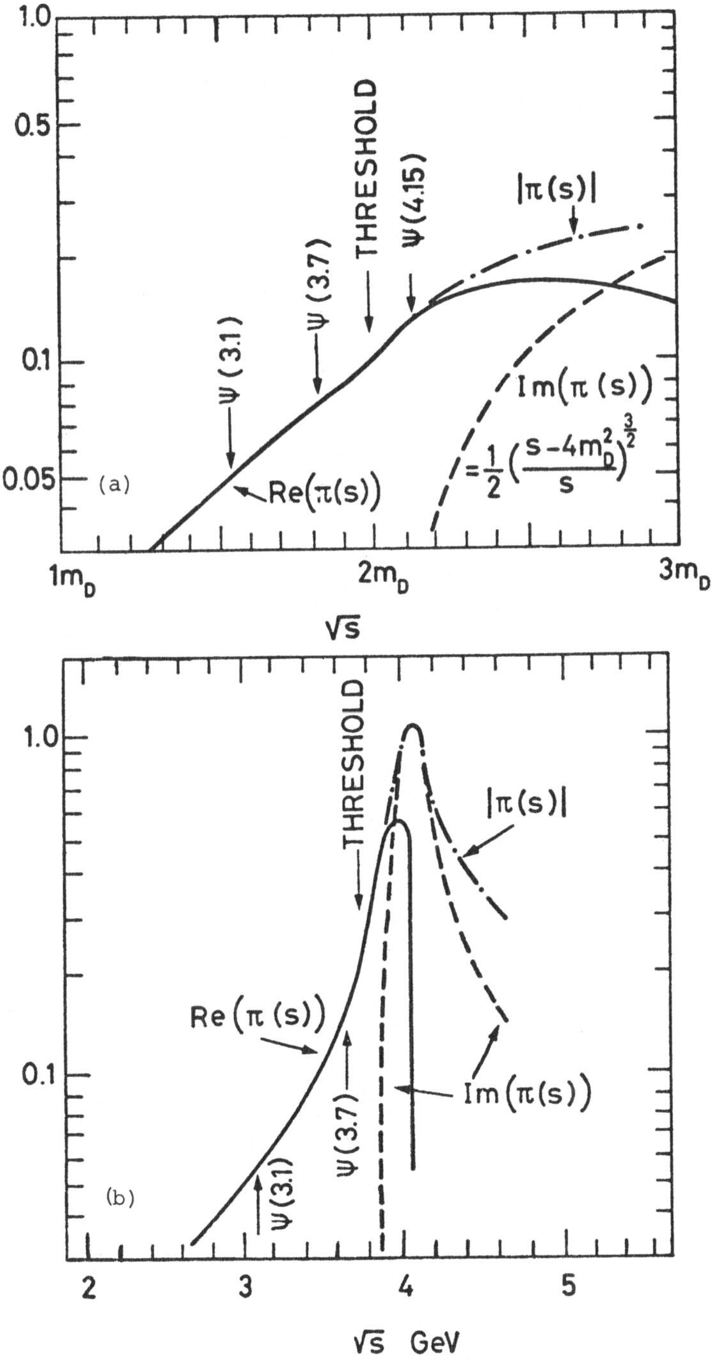

Figure 78 a,b: a) $\pi(q^2)$ for a point-coupling $\psi \to D\bar{D}$

b) $\pi(q^2)$ computed via dispersion relation supposing a resonance in Im $\pi(q^2)$ just above threshold (from Ref.118)

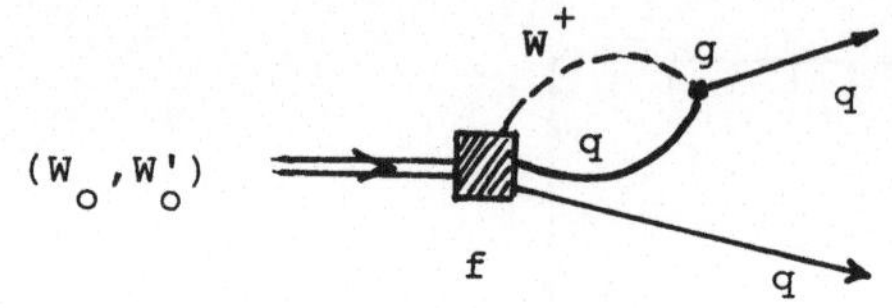

(a)

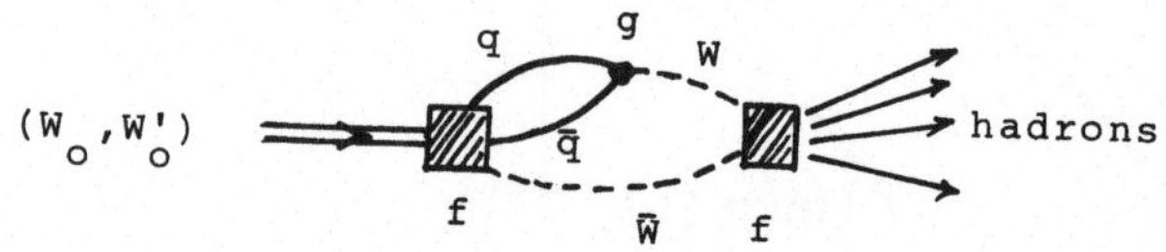

(b)

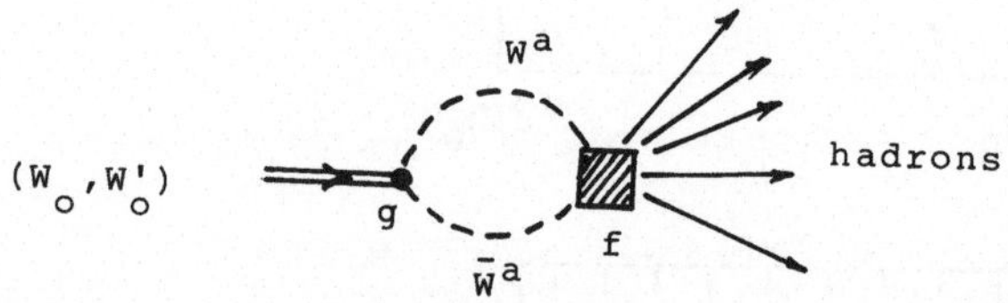

(c)

Figure 81a,b,c: Hadron production in the strong W-pair model

Lecture Notes in Physics

Bisher erschienen/Already published

Vol. 1: J. C. Erdmann, Wärmeleitung in Kristallen, theoretische Grundlagen und fortgeschrittene experimentelle Methoden. II, 283 Seiten. 1969.

Vol. 2: K. Hepp, Théorie de la renormalisation. III, 215 pages. 1969.

Vol. 3: A. Martin, Scattering Theory: Unitarity, Analyticity and Crossing. IV, 125 pages. 1969.

Vol. 4: G. Ludwig, Deutung des Begriffs „physikalische Theorie" und axiomatische Grundlegung der Hilbertraumstruktur der Quantenmechanik durch Hauptsätze des Messens. 1970. Vergriffen.

Vol. 5: M. Schaaf, The Reduction of the Product of Two Irreducible Unitary Representations of the Proper Orthochronous Quantum-mechanical Poincare Group. IV, 120 pages. 1970.

Vol. 6: Group Representations in Mathematics and Physics. Edited by V. Bargmann. V, 340 pages. 1970.

Vol. 7: R. Balescu, J. L. Lebowitz, I. Prigogine, P. Résibois, Z. W. Salsburg, Lectures in Statistical Physics. V, 181 pages. 1971.

Vol. 8· Proceedings of the Second International Conference on Numerical Methods in Fluid Dynamics. Edited by M. Holt. 1971. Out of print.

Vol. 9: D. W. Robinson, The Thermodynamic Pressure in Quantum Statistical Mechanics. V, 115 pages. 1971.

Vol. 10: J. M. Stewart, Non-Equilibrium-Relativistic Kinetic Theory. III, 113 pages. 1971.

Vol. 11: O. Steinmann, Perturbation Expansions in Axiomatic Field Theory. III, 126 pages. 1971.

Vol. 12: Statistical Models and Turbulence. Edited by C. Van Atta and M. Rosenblatt. Reprint of the First Edition. VIII, 492 pages. 1975.

Vol. 13: M. Ryan, Hamiltonian Cosmology. VII, 169 pages. 1972.

Vol. 14: Methods of Local and Global Differential Geometry in General Relativity. Edited by D. Farnsworth, J. Fink, J. Porter and A. Thompson. V, 188 pages.

Vol. 15: M. Fierz, Vorlesungen zur Entwicklungsgeschichte der Mechanik. V, 97 Seiten. 1972.

Vol. 16: H.-O. Georgii, Phasenübergang 1. Art bei Gittergasmodellen. IX, 167 Seiten. 1972.

Vol. 17: Strong Interaction Physics. Edited by W. Rühl and A. Vancura. V, 405 pages. 1973.

Vol. 18: Proceedings of the Third International Conference on Numerical Methods in Fluid Mechanics, Vol. I. Edited by H. Cabannes and R. Temam. VII, 186 pages. 1973.

Vol. 19: Proceedings of the Third International Conference on Nemerical Methods in Fluid Mechanics, Vol. II. Edited by H. Cabannes and R. Temam. VII, 275 pages. 1973.

Vol. 20: Statistical Mechanics and Mathematical Problems. Edited by A. Lenard. VIII, 247 pages. 1973.

Vol. 21: Optimization and Stability Problems in Continuum Mechanics. Edited by P. K. C. Wang. V, 94 pages. 1973.

Vol. 22: Proceedings of the Europhysics Study Conference on Intermediate Processes in Nuclear Reactions. Edited by N. Cindro, P. Kulišić and Th. Mayer-Kuckuk. XIV, 329 pages. 1973.

Vol. 23: Nuclear Structure Physics. Proceedings 1973. Edited by U. Smilansky, I. Talmi, and H. A. Weidenmüller. XII, 296 pages. 1973.

Vol. 24: R. F. Snipes, Statistical Mechanical Theory of the Electrolytic Transport of Non-electrolytes. V, 210 pages. 1973.

Vol. 25: Constructive Quantum Field Theory. The 1973 "Ettore Majorana" International School of Mathematical Physics. Edited by G. Velo and A. Wightman. III, 331 pages. 1973.

Vol. 26: A. Hubert, Theorie der Domänenwände in geordneten Medien. XII, 377 Seiten. 1974.

Vol. 27: R. K. Zeytounian, Notes sur les Ecoulements Rotationnels de Fluides Parfaits. XIII, 407 pages. 1974.

Vol. 28: Lectures in Statistical Physics. Edited by W. C. Schieve and J. S. Turner. V. 342 pages. 1974.

Vol. 29: Foundations of Quantum Mechanics and Ordered Linear Spaces. Advanced Study Institute, Marburg 1973. Edited by A. Hartkämper and H. Neumann. VI, 355 pages. 1974.

Vol. 30: Polarization Nuclear Physics. Proceedings 1973. Edited by D. Fick. IX, 292 pages. 1974.

Vol. 31: Transport Phenomena. Sitges International Schools of Statistical Mechanics, June 1974. Edited by G. Kirczenow and J. Marro. XIV, 517 pages. 1974.

Vol. 32: Particles, Quantum Fields and Statistical Mechanics. Proceedings 1973. Edited by M. Alexanian and A. Zepeda. V, 132 pages. 1975.

Vol. 33: Classical and Quantum Mechanical Aspects of Heavy Ion Collisions. Proceedings 1974. Edited by H. L. Harney, P. Braun-Munzinger, and C. K. Gelbke. VII, 311 pages. 1975.

Vol. 34: One-Dimensional Conductors. GPS Summer School Proceedings, 1974. Edited by H. G. Schuster. VII, 371 pages. 1975.

Vol. 35: Proceedings of the Fourth International Conference on Numerical Methods in Fluid Dynamics. June 24–28, 1974, University of Colorado. Edited by R. D. Richtmyer. V, 457 pages. 1975.

Vol. 36: R. Gatignol, Théorie Cinétique des Gaz à Répartition Discrète de Vitesses. II, 219 pages. 1975.

Vol. 37: Trends in Elementary Particle Theory. Proceedings 1974. Edited by H. Rollnik and K. Dietz. V, 472 pages. 1975.

Vol. 38: Dynamical Systems, Theory and Applications. Proceedings 1974. Edited by J. Moser. VI, 624 pages. 1975.

Vol. 39: International Symposium on Mathematical Problems in Theoretical Physics. Proceedings 1975. Edited by H. Araki. XII, 562 pages. 1975.

Vol. 40: Effective Interactions and Operators in Nuclei. Proceedings 1975. Edited by B. R. Barrett. XII, 339 pages. 1975.

Vol. 41: Progress in Numerical Fluid Dynamics. Proceedings 1974. Edited by H. J. Wirz. V, 471 pages. 1975.

Vol. 42: H II Regions and Related Topics. Proceedings 1975. Edited by D. Downes and T. L. Wilson. XII, 488 pages. 1975.

Vol. 43: Laser Spectroscopy. Proceedings 1975. Edited by S. Haroche, J. C. Pebay-Peyroula, T. W. Hänsch, and S. E. Harris. X, 466 pages. 1975.

Vol. 44: R. A. Breuer, Gravitational Perturbation Theory and Synchrotron Radiation. VI, 196 pages. 1975.

Vol. 45: Dynamical Concepts on Scaling Violation and the New Resonances in e^+e^- Annihilation. Edited by B. Humpert. VII, 248 pages. 1975.

Selected Issues from

Lecture Notes in Mathematics